MULTIVARIATE NOMINAL SCALE ANALYSIS

a report on a new analysis technique and a computer program

Frank M. Andrews
Robert C. Messenger

Survey Research Center • Institute for Social Research
The University of Michigan
Ann Arbor, Michigan
1973

ISR Code No. 3462

Library of Congress Catalog Card No. 72-619721
ISBN 0-87944-134-8 paperbound
ISBN 0-87944-135-6 clothbound

Published by the Institute for Social Research
The University of Michigan, Ann Arbor, Michigan 48106

Manufactured in the United States of America
Cover Design by Lena Behnke

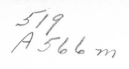
TABLE OF CONTENTS

Appendices

LIST OF TABLES AND FIGURES

Page

Chapter 1

INTRODUCTION

The multivariate analysis of attribute data--sometimes known as nominally scaled data--presents major problems to the data analyst. Few
methods are adequate for handling such data when one wishes to analyze
the relationships between one dependent variable and two or more independent variables. None of the existing methods is fully satisfactory
for analyses designed to explore the interrelationships of theoretical
concepts (in contrast to making a prediction), based on substantial numbers
of independent variables (e.g., 3-50) at various levels of measurement--
nominal, ordinal, and interval, and using moderate numbers of cases (e.g.,
300-6000). Such analysis problems, however, arise in many types of social
survey research and also in certain other fields.

The approach and computer program described in this monograph provides a new technique for analyzing such data. We have called it MNA,
signifying that it is for Multivariate Nominal scale Analysis. Of course
MNA has restrictions of its own, the major statistical one being that an
additive model appropriately describes the data being analyzed. This restriction, plus certain others, are more fully discussed later in the
monograph.*

As a new approach, MNA is relatively untested either by the statistical profession or by data analysts. Nevertheless, MNA seems sufficiently

---

*Another new method for analyzing nominal scales is also being
developed at the Institute for Social Research by James Morgan
and Robert Messenger. This latter approach is called THAID, is
briefly discussed in Chapter 4 and fully described in another
ISR monograph (Morgan and Messenger, 1973). THAID provides a
useful complement to MNA because of its non-additive model and
structure-searching orientation.

promising to present in its current form. We hope other users will join with us in helping to determine its worth and in developing refinements to it.

What we have described as a "new technique" rests on some older practices. It has been customary, when analyzing a dichotomous dependent variable, to use ordinary multiple regression and interpret the problem as predicting a probability. If the dependent variable contains three or more categories, membership in each can be treated as a dichotomous dependent variable and a set of such regressions run. It turns out that there are some convenient properties of the coefficients of a set of such regressions if one uses the identical predictors in all of them and dichotomous (dummy) predictors. What is described in this monograph is a program with simple and efficient input and output, designed to compute such sets of regressions and to provide some useful summary statistics.

Chapter 2 of this monograph discusses the kinds of problems to which MNA is applicable and gives an extended example of an MNA analysis. Chapter 3 provides a complete mathematical description of the computer program. Chapter 4 contrasts MNA with various other techniques and proposals for the multivariate analysis of attribute data. Chapter 5 details how one goes about using the MNA program, including a list of limitations. Chapter 6 discusses some unsolved problems relevant to MNA. The monograph concludes with a set of appendices giving complete set-up instructions, a macro flow chart of the computer program, a listing of some sample output, and information about how one may obtain and adapt the program for other computing installations.

It is a pleasure to acknowledge the help received in the development of MNA. Elizabeth Baker did much of the actual computer programming with marvelous speed and accuracy. Patrick O'Malley, Terrence Davidson, Jon Dickinson and Charles Cowen were users of MNA in its earliest forms. We have learned much from the examples they have contributed. Raymond Faith, Graham Kalton, Leslie Kish, Gary Koch, James Morgan, Judith Rattenbury, and Roger Wright provided helpful criticism of our ideas. Bonnie Brauer, Marita DiLorenzi, Nancy Mayer and Ellen Bronson all participated in typing various drafts of this monograph, which has been improved by suggestions from Laura Klem.

Financial support for the development of MNA came mainly from National Science Foundation grant #GI-29904, a general capability building grant to the Institute for Social Research.

Chapter 2

THE NATURE OF THE PROBLEM

2.0 Summary

This chapter outlines five general questions addressed by multivariate analysis and presents an extended example showing how MNA generates answers to each. The chapter serves as an introduction to the output obtainable from this technique, and a detailed guide to its interpretation. It also describes the kinds of data MNA is intended to handle.

2.1 The Problem

A general goal of multivariate data analysis is to understand how a dependent variable is affected by a set of independent variables. There are at least five general questions that can be posed: (1) Taken all together, how well do the independent variables explain the variability in the dependent variable? (2) What is the relationship of a particular independent variable to the dependent variable after statistically holding constant all other independent variables? (3) What is the marginal usefulness of a particular independent variable in explaining the dependent variable over and above what all other independent variables can explain? (4) Taking into account a particular object's (e.g., person's) scores on the independent variables, what score should we predict for it on the dependent variable? (5) By how much does each case actually deviate from the prediction made for it?

Multiple regression, together with multiple and partial correlation, provide classical answers to these questions when data meet the assumptions required of these techniques (interval scale measurement for all variables,* homoscedasticity, and linear additive relationships among all variables).

Multiple classification analysis (MCA)** can provide useful answers when the independent variables are not measured on interval scales and/or when relationships are not linear. MCA, however, still assumes interval measurement of the dependent variable, a normal distribution of residuals, and that effects operate additively.

Multiple discriminant function analysis represents still another technique for answering these questions. While it will accept categorical measurement of the dependent variable, it assumes that all independent variables are measured as interval scales and have linear additive effects with respect to the dependent variable. The multivariate analysis technique described in the present document--Multivariate Nominal scale Analysis (MNA)--is designed to handle problems where: (a) the dependent variable is a set of mutually exclusive categories--i.e., a nominal scale, and (b) where the independent variables may be measured at any level of measurement, including nominal measurement, and (c) where any form or pattern of relationship may exist (i) between any independent variable and the dependent variable, and (ii) between any pair of independent variables. While

---

*Four levels of measurement have been distinguished by Stevens (1946): nominal, ordinal, interval, and ratio. Nominal scales consist of sets of mutually exclusive categories--e.g., Fords, Chevrolets, Volkswagens, others. Ordinal scales are sets of mutually exclusive categories which have an order to them--e.g., low-priced cars, medium-priced cars, and high-priced cars. Interval scales (and also ratio scales) are sets of categories which have equal distances between them--e.g., prices expressed in dollars, temperatures expressed in degrees, distance expressed in miles, etc. (In this document we shall not need to distinguish between interval and ratio scales.)

**Andrews, Morgan, and Sonquist, 1967.

designed for use with nominal dependent variables, MNA may also be applied to an ordinal or interval dependent variable so long as the number of categories is small. As with the other multivariate techniques mentioned above, MNA assumes an additive model, an assumption discussed further in sections 5.2d and 6.1 of this monograph.*

Nominally scaled dependent variables are often encountered in the social sciences, and we suspect that if their analysis presented fewer problems than it has in the past, they would be even more common. Examples include orientations toward the use of violence; political party supported**; forms of contraception used; job changes experienced in the past five years; type of illness; product preferences; and many, many more. Given any one of the above as a dependent variable, it would not be unreasonable to want to understand how it was related to a set of independent variables which might include race, age group, education, region of the country, as well as the strength of certain attitudes, types of personalities, etc. No traditional multivariate analysis method can handle such data; MNA can --with the degree of adequacy depending on how well an additive model matches what is true of the real world.

---

*Since the assumption of additivity is the major restriction on the use of MNA, it may be useful to discuss it briefly here for readers not familiar with the term. What is implied is that the effect (on the probability of membership in categories of the dependent variable) of being a member of a particular category of one independent variable is the same for all people, regardless of their classification on other independent variables. When a simple additive model applies to a set of data, there is no "interaction" between two or more predictor variables and the dependent variable.

**Much multivariate analysis of data about American political parties has been possible because there happen to be only two major parties in the United States. When there are three or more parties of interest, traditional analytic methods become less useful.

2.2 An Example of Applying MNA

Since the insights which led to the development of MNA derived from an attempt to understand the bases of Americans' orientations toward the use of violence, we shall cite an example in this area to illustrate the application of MNA.* The emphasis here will be interpretive and not mathematical. A detailed mathematical description of MNA appears as Chapter 3.

Approximately 1300 male respondents to a national survey were classified into one of five violence-orientation types according to their expressed attitudes.** These types were named pacifists, anarchists, vigilantes, warriors, and a middle-of-the-road type known as intermediates. This five category typology is the dependent variable. The three independent variables of this example describe some aspects of each respondent's background--his race, the region of the country where he grew up, and his education. Applying MNA to these data produced the results shown in Table 2.1.***

2.2a Overall distribution.
The first point of interest is the overall percentage distribution of the sample across the several categories of the dependent variable. We observe that 25.0% were classified as "pacifists," 15.4% as "intermediates," etc. The modal category, as the groups were defined, was "vigilantes" which included 29.7% of the sample. (Identifying the size and location of the mode is interesting in the perspective of

---

*The example is chosen for its heuristic value; the relationships are rather weak and the predictive power small. Of course, this is a function of the data and not of MNA, which will indicate strong additive relationships when they are present.

**Full details about the data appear in Blumenthal, Kahn, Andrews, & Head, 1972. The book does not report results of applying MNA to the data, however, because MNA was developed later.

***Appendix C provides a listing of the actual computer output from which Table 2.1 was constructed.

Table 2.1

Major Output from an MNA Analysis*

	Pacifist	Inter-mediate	Warrior	Anarchist	Vigilante	Total
Overall percent	25.0	15.4	15.4	14.4	29.7	100

R^2 = .040

Θ = .364

R-squared	.043	.010	.044	.061	.043	

Race

η^2_{race} = .022

θ_{race} = .330

	Pacifist	Inter-mediate	Warrior	Anarchist	Vigilante	Total
Eta-squared	.005	.003	.024	.047	.033	
Beta-squared	.000	.006	.014	.056	.038	
White (87.9%; N = 949)						
Percent	26.1	16.1	13.4	11.6	32.8	100
Coefficient	0.3	1.0	-1.5	-3.1	3.3	0
Adjusted percent	25.3	16.5	13.9	11.3	33.0	100
Black (12.1%; N = 287)						
Percent	17.0	10.4	30.5	34.9	7.2	100
Coefficient	-2.0	-7.6	11.3	22.4	-24.1	0
Adjusted percent	23.0	7.9	26.8	36.8	5.6	100

Education

$\eta^2_{education}$ = .012

$\theta_{education}$ = .317

	Pacifist	Inter-mediate	Warrior	Anarchist	Vigilante	Total
Eta-squared	.026	.003	.020	.005	.005	
Beta-squared	.021	.004	.012	.008	.007	
>7 grades (4.6%; N = 70)						
Percent	15.0	19.2	32.5	10.8	22.5	100
Coefficient	-5.2	3.6	11.9	-7.5	-2.8	0
Adjusted percent	19.8	19.0	27.3	6.9	26.9	100
7-11 grades (33.9%; N = 423)						
Percent	17.9	17.1	17.1	15.4	32.4	100
Coefficient	-6.8	1.8	0.9	0.4	3.6	0
Adjusted percent	13.3	17.2	16.4	14.8	33.3	100

(continued)

Table 2.1 (continued)

	Pacifist	Inter-mediate	Warrior	Anarchist	Vigilante	Total
Completed high school (31.9%; N = 393)						
Percent	27.0	14.7	16.8	11.6	29.9	100
Coefficient	1.4	-0.7	2.0	-2.2	-0.4	0
Adjusted percent	26.4	14.7	17.4	12.2	29.3	100
College (25.1%; N = 299)						
Percent	30.4	14.7	10.1	15.9	29.0	100
Coefficient	5.0	-0.7	-4.6	1.8	-1.4	0
Adjusted percent	30.0	14.7	10.8	16.2	28.3	100
Graduate degree (4.5%; N = 51)						
Percent	45.4	8.4	5.9	21.8	18.5	100
Coefficient	18.5	-7.9	-7.3	10.2	-13.5	0
Adjusted percent	43.5	7.6	8.1	24.6	16.2	100

Region

η^2_{region} = .011

θ_{region} = .307

	Pacifist	Inter-mediate	Warrior	Anarchist	Vigilante	Total
Eta-squared	.022	.002	.018	.006	.004	
Beta-squared	.016	.004	.007	.006	.002	
New England (6.2%; N = 69)						
Percent	25.6	13.4	11.0	12.2	37.8	100
Coefficient	0.6	-2.7	-3.3	-0.4	5.7	0
Adjusted percent	25.7	12.8	12.1	14.0	35.4	100
Middle Atlantic (19.2%; N = 220)						
Percent	30.7	16.6	13.5	11.1	28.1	100
Coefficient	5.2	1.1	-1.1	-2.6	-2.6	0
Adjusted percent	30.2	16.5	14.3	11.8	27.1	100
East North Central (20.7%; N = 250)						
Percent	22.3	15.0	14.8	18.5	29.3	100
Coefficient	-3.8	-0.6	0.6	5.1	-1.4	0
Adjusted percent	21.3	14.9	16.0	19.5	28.3	100
West North Central & Mountain (15.3%; N = 166)						
Percent	34.2	13.4	8.7	12.9	30.8	100
Coefficient	7.8	-2.5	-4.5	0.6	-1.4	0
Adjusted percent	32.8	13.0	10.9	15.0	28.3	100
South & Border (31.4%; N = 458)						
Percent	17.5	17.1	22.1	15.7	27.6	100
Coefficient	-5.8	2.4	3.6	-1.9	1.7	0
Adjusted percent	19.2	17.8	19.1	12.5	31.3	100

(continued)

Table 2.1 (continued)

	Pacifist	Inter-mediate	Warrior	Anarchist	Vigilante	Total
Pacific (7.2%; N = 73)						
Percent	30.5	12.1	11.6	11.1	34.7	100
Coefficient	5.2	−4.2	−2.0	−0.5	1.6	0
Adjusted percent	30.2	11.2	13.4	13.9	31.3	100
Proportion classed correctly	.406	.000	.027	.279	.736	

Classification Matrix

Actual	Predicted					
	Pacifist	Intermediate	Warrior	Anarchist	Vigilante	Total
Pacifist	40.6%	0.0%	1.5%	6.7%	51.2%	100%
Intermediate	22.6	0.0	1.7	6.4	69.3	100%
Warrior	13.3	0.0	2.7	21.1	62.9	100%
Anarchist	22.4	0.0	1.3	27.9	48.4	100%
Vigilante	23.5	0.0	0.6	2.3	73.6	100%

---

*In addition to the statistics shown in this table, MNA outputs certain other figures (mainly weighted number of cases). Appendix C reproduces the actual output on which this table is based.

prediction because it shows that even if one knew nothing about the respondents, one could predict each would be a vigilante and be correct 29.7%
of the time. Relationships of the independent variables to the dependent
variable act to increase predictability above this 29.7% level.)

2.2b Multiple relationships. Another point of interest is the strength of
relationship between the independent variables taken together as a set and
the dependent variable. This is shown in two ways by MNA. One may note
in Table 2.1 that the generalized squared multiple correlation, R^2, is .04,
(equivalent to a multiple correlation of .20). This is roughly interpretable as having explained 4% of the "variance" in the dependent variable.*

Further analytic insight may be gained by examining the category-
specific R-squares. As shown in Table 2.1, the highest value is .061 (for
"anarchists") and the lowest is .010 (for the "intermediate" group). These
indicate that the anarchist category (i.e., whether or not a respondent
was an "anarchist") was best predicted by the independent variables used
in this analysis, and that the "intermediate" category was least well predicted.

One may also note that the multivariate Theta (denoted in Table 2.1 as Θ
and described more fully in Chapter 3) is .364--indicating that 36.4% of
the cases could be correctly classified after taking into account each
respondent's scores on the three independent variables. By comparing this
.364 with the mode of the overall percentage distribution (29.7%) one can

---

*The concept of "variance," when applied to a nominally scaled
dependent variable, is a subtle one. As detailed in Chapter 3,
the generalized R^2 is actually a variance-weighted average of the
R^2s which result from separate analyses of each category of the
dependent variable when each category is treated as a "dummy
variable."

see that use of these predictors in an additive model produces a gain in accuracy of prediction of 6.7 percentage points over what was achievable without taking these variables into account (36.4 - 29.7 = 6.7).

The generalized multiple correlation (squared) and the multivariate Theta are the statistics by which MNA answers the first general question of multivariate analysis--joint explanatory power--listed above in Section 2.1.*

Having examined the overall percentage distribution and the strength of the multiple relationship, one turns next to consider each independent variable individually. MNA produces a variety of statistics for each variable.

2.2c Variable specific relationships--summary statistics. The first listed variable is race, with two categories--whites and blacks. One may see that a total of 949 respondents were whites and that this was 87.9% of the weighted cases being analyzed.** Similarly, there were 287 blacks, 12.1% of the weighted total.

The generalized eta square, η^2_{race} and the bivariate Theta, θ_{race},*** provide two alternative ways of measuring the strength of the simple bivariate relationship between Race and Orientation-toward-violence. Note that these statistics perform the same function in the bivariate situation as were performed for the multivariate situation by the R^2 and Θ statistics discussed above.

---

*For comments about the different perspectives represented by R^2 and Θ see Sections 2.2g, 3.5, and 3.6 and the accompanying footnotes.

**The computer program accepts an optional "weight" variable for use with weighted samples. If a weight variable is used, all statistics (except the number of cases) are based on computations which include these weights.

***Chapter 3 describes these statistics.

The generalized eta square is .022, indicating a mild association between these two variables. As before, analytic insight may be gained by
examining the category-specific eta squares: one sees that Race is most
useful for distinguishing "anarchists" from everybody else, and that it
is virtually useless for distinguishing "pacifists" or "intermediates" from
others.

The bivariate Theta, θ_{Race}, indicates that knowing a respondent's race
would permit correct prediction of about one third (.330) of the cases.
This represents a modest gain over the 29.7% that could have been predicted
correctly without knowing anything about the respondents.

Since the generalized eta square for race is not as high as the generalized R^2, nor the bivariate Theta for race as high as the multivariate
Theta, one concludes that one or both of the other independent variables
in the analysis--region and education--contributed something to predicting
the dependent variable over and above the predictive power of race alone.

The program also puts out a series of statistics labelled "Beta
Square." These provide an indication of the importance of Race as a predictor of each category of the dependent variable when holding constant
all other independent variables.* The Beta statistic is the means by which
MNA provides a summary answer to the second general question of multivariate
analysis (as listed in Section 2.1). However Beta is only a summary answer,
and more detailed answers are provided by the "coefficients" and "adjusted
percents" in the following sub-section.

---

*We regard the Beta square as an experimental statistic whose
precise interpretation is open to further investigation. We
discuss it further in Section 6.5b. Section 6.5a distinguishes
three different criteria for assessing predictor importance; Beta
is relevant to only one.

2.2d Variable-specific relationships--detailed statistics. Once the summary statistics have been examined for Race, one may wish to examine in detail how each race category is related to each type of violence orientation.
Three sets of figures in Table 2.1 are useful for this.

First, the rows labelled "Percent" show the percentage distribution
of whites (or blacks) across the violence types. Comparing the two rows of
percents shows how whites and blacks differ with respect to violence type
and is a more detailed way of examining the bivariate relationships summarized by the generalized eta square and the bivariate Theta.

(One of the results shown in Table 2.1 is that blacks were less likely
to fall in the "pacifist" category than were whites--17.0% versus 26.1%.
We call attention to this here, because it stands in instructive contrast
to the adjusted percentages, to be discussed below.)

The rows labelled "coefficients" are of interest in a multivariate
perspective (in contrast to the bivariate perspective relevant for the "percent" row). The coefficients show the "effects" of membership in the particular category of the independent variable on the likelihood of membership in each category of the dependent variable. These effects are the
core of the additive model on which MNA is based. These are the figures
which are literally added together (across the several independent variables)
to predict the individual's score on the dependent variable. (The expected
probability for any individual is the sum of the coefficients which pertain
to that individual plus the base likelihood--shown in the row labelled
"overall percent.")

It is important to recognize that the coefficients take into account
any relationships that may be present between the various independent variables and between each independent variable and the dependent variable.

Thus they can be interpreted as indicating the gain or loss in likelihood after "holding constant" all other independent variables. Another way of saying this is that the coefficients indicate what the effect of a particular category (e.g., "black" or "white") would be if the members of this category were distributed as in the general population with respect to all other predictor variables.

Reference to the numbers in Table 2.1 may help to clarify this last statement. It was noted earlier that in a bivariate sense blacks were less likely to be "pacifists" than were whites. However the coefficients show that the bivariate difference is largely (but not completely) removed when one holds constant region and education, the other independent variables in this analysis. (Blacks are only 2 percentage points less likely than the average American to be pacifists, and whites only 0.3 points more likely, after holding constant region and education.) It would appear that part of the racial difference observed in the bivariate percentages can be attributed to the fact that blacks tend to have less education than whites (a fact not shown in Table 2.1), and the fact that pacifism is positively related to education (shown by the coefficents for education in Table 2.1).

The third row of detail figures, labelled "adjusted percents," shows, in each cell, the sum of the coefficient for that cell plus the relevant base likelihood ("overall percent"). Thus in the case of black pacifists, one notes that the "adjusted percent" is 23.0 (-2.0 + 25.0 = 23.0). That this 23.0% is not much different from the overall percent of 25.0% is another way of observing that race makes little difference with respect to pacifism after holding constant region and education.

We will not further review the specific findings shown in Table 2.1.

Suffice it to note that similarly detailed study of statistics shown for the other independent variables would be undertaken by the analyst interested in the results.

2.2e Scanning strategies for detailed statistics. In examining a large number of detail statistics, two phenomena would be of particular interest to the analyst: (1) the occurrence of large coefficients, and (2) the occurrence of large differences between the "percents" and the "adjusted percents." When one or both of these occur, something "interesting" is indicated.

If the independent variable happens to be an ordinal scale (as is Education in the example), one additional phenomenon may be of interest-- the occurrence of monotonic change between successive coefficients or monotonic change of percentages across the categories of the independent variable. An approximation of this occurs in the way education affects the likelihood of being a pacifist. One may note that the percents, the coefficients, and the adjusted percents all tend to become larger (or more positive, in the case of the coefficients) as education increases.

2.2f Marginal predictive power. The third general question addressed by multivariate analysis--as listed in Section 2.1--concerns the marginal usefulness of each predictor over and above all other predictors. No statistic from a single MNA run provides an answer to this question. However, this can be assessed by making two runs of MNA. The two runs must be identical except for the omission from one of them of the predictor whose marginal power is to be assessed. The difference in the joint predictive power of the included predictors (measured by R^2 or Θ) is the marginal power of the omitted predictor. (See Section 6.5a for additional comments.)

In the present example, if one wished to know how useful Race was in predicting Orientation-toward-violence, <u>over and above</u> what could be predicted on the basis of Education and Region of childhood, one would make a second MNA run using only the latter two variables as predictors. The difference between the R^2 or Θ obtained in this second run and that obtained in the first run--as shown in Table 2.1--would be a measure of the marginal power of Race.

2.2g Predicted scores--forecasts and the proportion classed correctly. As noted at the beginning of this chapter, one of the functions of multivariate analysis is to predict scores on a dependent variable according to the pattern of scores shown by an individual case (e.g., person) on the independent variables. MNA does this in two stages.

For any case a "forecast" can be derived. The forecast shows the likelihood of that individual falling into each category of the dependent variable--i.e., it consists of a set of probabilities. Each probability is computed by summing, separately for each category of the dependent variable, the coefficients relevant to that particular case (and adding in the overall percent). If the case has a pattern of characteristics on the independent variables identical to one that occurs in the data which were analyzed, and if an additive model correctly represents the data, the forecast numbers will all lie in the range 0-100, will sum to 100, and can be interpreted as probabilities (after an appropriate shift of the decimal point).*

Using the data in Table 2.1 we can construct the forecast for any actual (or hypothetical) case. Assume we have a person who is white, who grew up in New England, and who has some college education but no graduate

---

*Section 3.4 provides an explanation as to why MNA has this convenient property.

degree. His forecast would be computed as shown in Table 2.2.

Table 2.2

Example of Forecast Computed for White, College Educated
New Englander Using Data from Table 2.1

Violence orientation:	Pacifist	Inter mediate	Warrior	Anarchist	Vigilante
Overall percents	25.0	15.4	15.4	14.4	29.7
Coeff: whites	0.3	1.0	-1.5	-3.1	3.3
Coeff: New England	0.6	-2.7	-3.3	-0.4	5.7
Coeff: college educ.	5.0	-0.7	-4.6	1.8	-1.4
Forecast	30.9	13.0	6.0	12.7	37.3

The forecast represents a <u>set</u> of predicted scores for each case and provides a way of arriving at a single predicted score for each case on the original dependent variable. One would predict a case to be in the dependent variable category for which the probability was highest. (Thus the person whose forecast is computed in Table 2.2 would be assigned to "vigilante.")*

MNA actually computes the forecast for each case in the data set, assigns the objects to categories of the dependent variable, and checks to see how well the predicted scores agree with the actual scores. Table 2.1 includes a row labelled "proportion classed correct" indicating the proportion of individuals actually located in each violence type who were cor-

---

*Although our description is in terms of predicting a category for a <u>single</u> case (e.g., person), one can also consider MNA as predicting probabilities for <u>classes</u> of cases, where classes have common characteristics on the independent variables. The Theta statistics are appropriate when one adopts the former perspective, and the eta, Beta, and R^2 statistics when one adopts the latter. R^2 is the proportion of variance (around the mean) explained by the predictor, and in the case of a dummy variable, as in MNA, the variance is $(p)(1-p)$--where P is the proportion of cases falling in one category of the dummy variable.

rectly predicted by MNA as being in that category. One may note a typical result--that accuracy of prediction varies substantially across the categories of the dependent variable. (As mentioned above, the multivariate Theta shows the proportion correctly classed for the data set as a whole. It can be thought of as a frequency-weighted average of the "proportion classed correct" figures.)

Table 2.1 also includes a Classification Matrix comparing the actual classifications on the dependent variable with the categories which were predicted by MNA. These data supplement the "Proportion classed correct" figures by indicating the nature of the errors made among cases that were not correctly classed. One may observe, for example, that of all the Pacifists, 40.6% were (correctly) predicted to be in the pacifist category, none were (incorrectly) predicted to be in the Intermediate category, 1.5% were (incorrectly) predicted to be Warriors, etc.

2.2h Some properties of MNA results. In interpreting the results of an MNA analysis it is useful to know several properties of the percentages and coefficients (discussed in greater detail in Chapter 3):

(a) Percentages--"overall percents," "percents," and "adjusted percents" always sum to 100% across the categories of the dependent variable.

(b) "Coefficients" associated with any category of any independent variable always sum to zero across the categories of the dependent variable.

(c) "Coefficients" associated with any category of the dependent variable always sum to zero across the categories of any independent variable when weighted by the sum-of-weighted-cases in the categories of the independent variable.

(d) The "forecast" for any individual involves a set of numbers which
will always sum to 100.

2.3 Residual Scores

Given a forecast for each individual in a data set, and a known actual
score, it is sometimes a useful analytic strategy to compute residual scores.
The residual scores are simply the differences between the actual situation--
scored as "probability" of 0 or 1--on each category of the dependent variable
and the MNA-generated probabilities. These residual-score variables can be
subjected to additional analysis in an attempt to explain variance which
could not be explained by the independent variables in the original set.
The residual-score variables are MNA's answer to the fifth general question
addressed by multivariate analysis.

There will be as many residual-score variables as there are categories
of the original dependent variable, these variables will be interval scales
which normally vary between +1 and -1, and will be analyzable by techniques such as Multiple Regression or Multiple Classification Analysis.
Sections 5.4 and 6.4 provide additional information about the computation
and use of residual scores and a brief example.

Chapter 3

MATHEMATICAL DESCRIPTION OF MNA

3.0 Summary

This chapter defines the mathematical model underlying MNA, discusses two important properties present in the model, and gives both the definition of and the rationale behind the set of summary statistics. Because of the parallels found in the Multiple Classification Analysis (MCA) monograph (Andrews, Morgan, and Sonquist, 1967) the notation was chosen as similar as possible. Further chapters will use the basic terms given below with definition of additional notation for specific local usage.

3.1 Basic Terms*

$A_{\ell m}$ $\equiv$ m^{th} transformed dummy predictor regression coefficient for ℓ^{th} dummy dependent variable.

$a_{ij\ell}$ $\equiv$ transformed dummy predictor regression coefficient of jth code of ith predictor for ℓ^{th} dummy dependent variable.

$B_{\ell m}$ $\equiv$ m^{th} dummy predictor regression coefficient for ℓ^{th} dummy dependent variable.

$b_{ij\ell}$ $\equiv$ dummy predictor regression coefficient of jth code of ith predictor for ℓ^{th} dummy dependent variable.

$\beta_{i\ell}^2 = \dfrac{\sum\limits_{j} w_{ij} a_{ij\ell}^2}{T_\ell}$ $\equiv$ beta-square for i^{th} predictor on ℓ^{th} dummy dependent variable.

---

*For convenience in derivations, two subscript conventions are used to identify the coefficients A,a,B and b: the ℓ subscript is used across all predictors while ij refers to code j within predictor i.

$$\beta_i^2 = \frac{\sum_\ell \sum_j w_{ij} a_{ij\ell}^2}{\sum_\ell T_\ell} \equiv \text{Generalized beta-squared for } i\text{th predictor.}$$

$C_i \equiv$ number of non-empty categories in the ith predictor.

$C = \sum_i C_i \equiv$ total number of non-empty predictor categories (also number of dummy predictors).

$\delta_{\alpha,\gamma} = \{ \begin{matrix} 1 \text{ if } \alpha=\gamma \\ 0 \text{ if } \alpha\neq\gamma \end{matrix} \equiv$ Kronecker delta function.

$E_\ell = \sum_k w_k (\hat{y}_{k\ell} - \bar{y}_\ell)^2 \equiv$ MNA explained sum of squares for ℓth dependent variable.

$G \equiv$ number of non-empty categories in the dependent variable.

$i \equiv$ predictor subscript.

$j \equiv$ predictor code subscript.

$k \equiv$ individual subscript.

$\ell \equiv$ dummy dependent variable subscript.

$m \equiv$ dummy predictor subscript.

$M_{ij} \equiv$ modal frequency of jth category of ith predictor (weighted).

$N \equiv$ number of individuals, observations or cases.

$\eta_{i\ell}^2 = \frac{V_{i\ell}}{T_\ell} \equiv$ one-way analysis of variance fraction of variance explained (eta-squared) on ℓth dummy dependent variable by ith predictor.

$\eta_i^2 = \frac{\sum_\ell V_{i\ell}}{\sum_\ell T_\ell} \equiv$ generalized eta-squared for ith predictor.

$P \equiv$ number of predictors.

$q_i \equiv$ transform constant for ith predictor.

$Q_{ij} \equiv$ Subset of individuals having jth code value on predictor i.

$r = C-P \equiv$ Number of dummy predictors in non-transformed coeff. model.

$R_\ell^2 = \dfrac{E_\ell}{T_\ell} \equiv$ regression fraction of variance explained for the ℓth dummy dependent variable.

$R_\ell^{2'} = 1 - (1 - R_\ell^2)\ \left(\dfrac{N-1}{N-1-C+P}\right)\ = R_\ell^2 - \dfrac{C-P}{N-1-C+P}\ (1-R_\ell^2) \equiv$ adjusted fraction of variance explained for the ℓth dummy dependent variable.

$\mathcal{R}^2 = \dfrac{\sum\limits_{\ell} E_\ell}{\sum\limits_{\ell} T_\ell} \equiv$ Generalized R-squared for the analysis.

$S_\ell^2 = \dfrac{\sum\limits_{k} w_k (y_{k\ell} - \bar{y}_\ell)^2}{\sum\limits_{k} w_k} \equiv$ sample variance of ℓth dummy dependent variable.

$T_\ell = \sum\limits_{k} w_k (y_{k\ell} - \bar{y}_\ell)^2 \equiv$ total sum of squares for ℓ^{th} dummy dependent variable.

$\theta_i = \dfrac{\sum\limits_{j} M_{ij}}{\sum\limits_{k} w_k} = \dfrac{\sum\limits_{j} w_{ij} \max\limits_{\ell} (\bar{y}_{ij\ell})}{\sum\limits_{k} w_k} \equiv$ Bivariate Theta for predictor i.

$\Theta = \dfrac{\sum\limits_{k} w_k \delta_{z_k, \hat{z}_k}}{\sum\limits_{k} w_k} \equiv$ Multivariate Theta for the analysis.

$V_{i\ell} = \sum\limits_{j} w_{ij} (\bar{y}_{ij\ell} - \bar{y}_\ell)^2 \equiv$ one-way analysis of variance explained sum of squares for the ℓth dummy dependent variable by the ith predictor.

$V =$ Variance-covariance matrix of dummy predictor variables.

w_k ≡ individual k's weight. (In all instances the weight vector, $[w_k]$, contains sampling weights arrived at prior to analysis and does not possess any analytical properties.)

$w_{ij} = \sum_{k \varepsilon Q_{ij}} w_k$ ≡ weight sum for j^{th} code of the ith predictor.

x_{km} ≡ m^{th} dummy predictor score for k^{th} individual.

X_{ik} ≡ i^{th} predictor score for k^{th} individual.

$$\bar{y}_\ell = \frac{\sum_k w_k y_{k\ell}}{\sum_k w_k}$$ ≡ sample mean for ℓ^{th} dummy dependent variable.

$$\bar{y}_{ij\ell} = \frac{\sum_{k \varepsilon Q_{ij}} w_k y_{k\ell}}{w_{ij}}$$ ≡ sample mean for ℓ^{th} dummy dependent variable for individuals in j^{th} code of ith predictor.

$\hat{y}_{k\ell}$ ≡ individual k's MNA prediction for ℓ^{th} dummy dependent variable.

$y_{k\ell}$ ≡ individual k's score on the ℓth dummy dependent variable.

$\hat{z}_k$ = $\ell: \max_\ell (\hat{y}_{k\ell})$ ≡ individual k's predicted category on the dependent variable.

z_k ≡ individual k's category on the dependent variable.

3.2 Regression System

The MNA program is based on the principle of repeated application of least squares dummy variable regression (Suits, 1957). In this approach the set of original predictor variables $\{X_1, X_2, \ldots, X_p\}$ is converted into a set of dummy predictor variables $\{x_1, x_2, \ldots, x_{c_1}, \ldots x_r\}$ by the following procedure. First, <u>every</u> non-empty code of each predictor is treated as a new dummy variable and assigned a value 1 when that code appears and a 0 otherwise. (In theory, an arbitrary choice of the two dummy variable codes could be made, as long as they are different, but 0,1 have convenient computational properties.)

As an example of the result of this step consider Table 3.1 giving values for the two trichotomous predictors X_1 and X_2 which are translated into dummy variable predictors $x_1, x_2, \ldots x_6$.

Table 3.1

<u>Dummy</u> <u>Predictor</u> <u>Conversion</u>

	Trichotomous Predictors		Dummy Variable Predictors					
	X_1	X_2	x_1	x_2	x_3	x_4	x_5	x_6
Case 1	2	1	0	1	0	1	0	0
Case 2	1	3	1	0	0	0	0	1
Case 3	3	2	0	0	1	0	1	0

In this table, we see that the data set of dummy predictors has 2 linear dependencies, one for each set of dummy predictors associated with an original predictor. This is expressed in either set by

$$x_i = 1 - \sum_{j \neq i} x_j . \tag{3-1}$$

Left as is this would lead to linear dependencies and a singular matrix in the least squares normal equations (3-5). The linear dependencies are removed by omitting one dummy predictor from each set (the choice is arbitrary since removal of any code from a set provides linear independence). The result of this is a set of r=C-P linearly independent "dummy-ized" predictors.

This set of r dummyized predictors is now applied successively to the complete set of G dummy dependent variables (note that G is the number of non-empty dependent variable codes, i.e. that no dummy variable has been omitted from the dependent variable set) to minimize the error sum of squares, which forms the least squares criterion, given by

$$ESS_\ell = \sum_k w_k (y_{k\ell} - \hat{y}_{k\ell})^2 \qquad (\ell=1,2,\ldots G) \qquad (3-2)$$

where

$$\hat{y}_{k\ell} = B_{\ell o} + B_{\ell 1} x_{k1} + B_{\ell 2} x_{k2} + \ldots + B_{\ell r} x_{kr} \qquad (\ell=1,2,\ldots G) \ . \qquad (3-3)$$

Taking partial derivatives of the ESS's with respect to the B coefficients and setting them equal to zero, i.e.,

$$
\begin{bmatrix}
\dfrac{\partial ESS_\ell}{\partial B_{\ell o}} \\[2em]
\dfrac{\partial ESS_\ell}{\partial B_{\ell 1}} \\[2em]
\vdots \\[1em]
\dfrac{\partial ESS_\ell}{\partial B_{\ell r}}
\end{bmatrix}
=
\begin{bmatrix}
0 \\[2em]
0 \\[2em]
\vdots \\[1em]
0
\end{bmatrix}
\qquad (\ell=1,2,\ldots G) \qquad (3-4)
$$

gives the G "normal equations" sets (Cooley and Lohnes, 1971):

$$
\begin{bmatrix}
\sum_k w_k y_{k\ell} \\[1em]
\sum_k w_k x_{k1} y_{k\ell} \\[1em]
\vdots \\[1em]
\sum_k w_k x_{kr} y_{k\ell}
\end{bmatrix}
=
\begin{bmatrix}
\sum_k w_k & \sum_k w_k x_{k1} \cdots \sum_k w_k x_{kr} \\[1em]
\sum_k w_k x_{k1} & \sum_k w_k x^2_{k1} \\[1em]
\vdots & \vdots \quad \vdots \quad \ddots \\[1em]
\sum_k w_k x_{kr} & \sum_k w_k x_{k1} x_{kr} \cdots \sum_k w_k x^2_{kr}
\end{bmatrix}
\begin{bmatrix}
B_{\ell 0} \\[1em]
B_{\ell 1} \\[1em]
\vdots \\[1em]
B_{\ell r}
\end{bmatrix}
. \quad (3\text{-}5)
$$

$$(\ell = 1, 2, \ldots G)$$

Solution of these G sets yields the B's for the <u>system</u> of predictive equations (3-3) and the set $\{\hat{y}_{k1}, \hat{y}_{k2}, \ldots, \hat{y}_{kG}\}$ is termed the "forecast" for individual k.

3.3 Transformed Regression Coefficients

It is possible to represent the predictive equations in an easier to interpret form while at the same time "putting back" the arbitrarily omitted codes by using the following transformation. First, let b_{ij} be the jth non-transformed dummy predictor coefficient for the ith predictor and $\bar{y}$ the grand mean in <u>one</u> regression with $b_{ij}=0$ for the omitted codes, b the constant term, and q a constant for the ith predictor. Now let a_{ij} be the jth transformed dummy predictor coefficient for the ith predictor such that

$$a_{ij} = b_{ij} + q_i \qquad \text{constrained by} \qquad (3\text{-}6)$$

$$\sum_j w_{ij} a_{ij} = 0 \qquad (i=1,2,\ldots,P) \quad . \qquad (3\text{-}7)$$

It follows then that

$$q_i = \frac{-\sum_j w_{ij} b_{ij}}{\sum_k w_k} \qquad \text{and} \qquad (3\text{-}8)$$

$$\sum_i q_i = b - \bar{y} \qquad \text{since} \qquad (3\text{-}9)$$

$$\sum_{ij} w_{ij} b_{ij} = \sum_k w_k (\hat{y} - b) \qquad \text{and} \qquad (3\text{-}10)$$

$$\sum_k w_k \hat{y}_k = \sum_k w_k \bar{y} \qquad ; \qquad \text{therefore} \qquad (3\text{-}11)$$

if we subtract $(B_{\ell o} - \bar{y}_\ell)$ from the predictive equations (3-3) and replace the B's by the A's according to (3-6) we have the transformed system

$$\hat{y}_\ell = \bar{y}_\ell + A_{\ell 1} x_1 + A_{\ell 2} x_2 + \ldots + A_{\ell C} x_C \qquad (\ell = 1,2\ldots G) \qquad (3\text{-}12)$$

that gives an _identical_ forecast $\{\hat{y}_{k1}, \hat{y}_{k2}, \ldots, \hat{y}_{kG}\}$ for all individuals k, has _grand_ _means_ $\{\bar{y}_1, \bar{y}_2, \ldots, \bar{y}_G\}$ for constant terms, and includes the _complete_ set of dummy predictors, as desired.

3.4 Properties of the Regression System

There are two properties of the system (3-12) that are worth noting. The first, termed the "unity-forecast" property, is stated mathematically by

$$\sum_\ell \hat{y}_{k\ell} = 1.0 \qquad \text{for all k} \quad . \qquad (3\text{-}13)$$

This property is convenient and adds support to an interpretation of the forecast as estimates of the G population proportions. The proof is as follows. Let (3-5) be denoted in matrix notation by

$$[y*_\ell] = [V][B_\ell] \qquad (\ell = 1, 2, \ldots, G) \quad . \qquad (3\text{-}14)$$

Since $\{y_{k1}, y_{k2}, \ldots, y_{kG}\}$ form a dummy variable set

$$y_{k\ell} = 1 - \sum_{n \neq \ell} y_{kn}, \qquad \text{for all } k, \qquad (3\text{-}15)$$

and

$$[y^*_\ell] = \begin{bmatrix} \sum_k w_k \\ \sum_k w_k x_{k1} \\ \vdots \\ \sum_k w_k x_{kr} \end{bmatrix} - \sum_{n \neq \ell} [y^*_n] . \qquad (3\text{-}16)$$

From (3-14),

$$[B_\ell] = [V]^{-1}[y^*_\ell] \qquad (3\text{-}17)$$

and thus, by substituting (3-16),

$$[B_\ell] = \begin{bmatrix} 1 \\ 0 \\ 0 \\ \vdots \\ 0 \end{bmatrix} - \sum_{n \neq \ell} [B_n] \qquad (3\text{-}18)$$

or

$$\sum_\ell [B_\ell] = \begin{bmatrix} 1 \\ 0 \\ 0 \\ \vdots \\ 0 \end{bmatrix} . \qquad (3\text{-}19)$$

Now, since

$$\sum_\ell q_{i\ell} = \sum_\ell \frac{-1}{\sum_k w_k} \sum_j w_{ij} B_{\ell ij} = \frac{-1}{\sum_k w_k} \sum_j w_{ij} \sum_\ell B_{\ell ij} = 0 \quad , \qquad (3\text{-}20)$$

by virtue of (3-6) and (3-19)

$$\sum_\ell [A_\ell] = \begin{bmatrix} 1 \\ 0 \\ 0 \\ \vdots \\ 0 \end{bmatrix} \qquad \text{also,} \qquad (3\text{-}21)$$

where $[A_\ell]$ is the ℓth transformed coeff. vector as in (3-12) with $A_{\ell o} \equiv \bar{y}_\ell$.

The second property is known as the "end-effect" and is stated simply by the fact that despite (3-13)

$$0 \nleq \hat{y}_{k\ell} \quad \text{and} \quad \hat{y}_{k\ell} \nleq 1.0, \qquad \text{in general.} \qquad (3\text{-}22)$$

A little reflection, however, shows that $\hat{y}_{k\ell}$ outside the (0,1) probability number range happens for only two reasons which are:

A.   There exists interaction in the sample.

B.   The $\hat{y}_{k\ell}$'s outside the (0,1) range are calculated on individuals

not having corresponding predictor code sets in the sample.

Reason A can be demonstrated by assuming no interaction in the sample. If no interaction is present then the $\hat{y}_{k\ell}$'s are the sample proportions from the marginal frequency distribution having the same predictor code set as individual k and therefore must lie in (0,1).

3.5 Bivariate Statistics

Two basic statistics are used to measure the strength of the bivariate relationship between the dependent variable and each predictor.*   These are:

---

*A third statistic, the generalized beta-square, β^2_i, is presently under consideration as an analogue to η^2_i after "adjustment" for other predictors. (See Chapter 6.)

(1) the one-way analysis of variance eta-squared statistic, $\eta^2_{i\ell}$, calculated for each dummy dependent variable and summarized in the generalized eta-square, η^2_i, discussed below; and (2) bivariate Theta, θ_i, used to measure strength of association using a criterion of correct placement in the dependent variable code.

The one-way analysis of variance eta-square statistic (Hays, 1963) measures the ability of the predictor to explain the variance of each dependent variable code dichotomized against all others. A natural generalization of this across all codes is the ratio of the sum of explained sums of squares to the sum of total sums of squares. That is,

$$\eta^2_i = \frac{\sum_\ell V_{i\ell}}{\sum_\ell T_\ell} , \qquad \text{also expressed as} \qquad (3\text{-}23)$$

$$\eta^2_i = \frac{\sum_\ell S^2_\ell \eta^2_{i\ell}}{\sum_\ell S^2_\ell} , \qquad (3\text{-}24)$$

the variance weighted average of $\eta^2_{i\ell}$'s. Note that in the special case when the dependent variable has two categories $S^2_1 = S^2_2$, $\eta^2_{i1} = \eta^2_{i2}$ and $\eta^2_i = \eta^2_{i1}$.

The bivariate Theta, θ_i, statistic (Messenger and Mandell, 1972) is a linear transformation of the Goodman and Kruskal Lambda statistic (Goodman and Kruskal, 1954), λ_i. It is thought to have more intuitive appeal than Lambda and fits naturally with its multivariate version discussed later. Theta is defined simply as the proportion of the sample correctly classed when using a prediction-to-the-mode strategy in the frequency distribution of each category of the predictor variable and is given by

$$\Theta_i = \frac{\sum_j w_{ij} \max_\ell (\bar{y}_{ij\ell})}{\sum_k w_k} \quad .$$ (3-25)

This is illustrated in Table 3.2 where the sample of 100 cases for dependent variable Z having 3 codes is cross-tabulated by predictor variable X also having 3 codes.

Table 3.2

Bivariate Theta Example

Z

		1	2	3	
	1	20	10	0	30
X	2	10	20	10	40
	3	0	10	20	30
		30	40	30	100

Let $\Theta.$ denote the proportion of the sample correctly classed when predicting to the mode of the univariate distribution of Z. Then if Θ_i denotes the proportion correctly classed when predicting to the appropriate marginal frequency mode, we have

$$\Theta. = \frac{40}{100} = .40$$ (3-26)

and
$$\Theta_i = \frac{20+20+20}{100} = .60 \quad .$$ (3-27)

λ_i is defined as the proportional reduction in errors given the predictor X_i's code. Therefore

$$\lambda_i = \frac{\Theta_i - \Theta_.}{1 - \Theta_.} = .33 \quad . \tag{3-28}$$

since $\frac{1}{1-\Theta_.}$ and $\Theta_i - \Theta_.$ are both >0, λ_i and Θ_i rank predictors identically.*

3.6 Multivariate Statistics

Two statistics are used to measure the multivariate strength of asso-ciation. These are the generalized squared multiple regression coefficient, $\mathcal{R}^2$, and the multivariate version of the Theta statistic, Θ. As in the gen-eralized eta-square statistic, the rationale behind the generalized squared multiple correlation coefficient was to combine the well known individual R_ℓ^2 statistics maintaining optimality and $(0,1)$ range in as simple and easy to interpret way as possible. The ratio of sums of explained sums of squares to total sums of squares resulted and gives

$$\mathcal{R}^2 = \frac{\sum_\ell E_\ell}{\sum_\ell T_\ell} = \frac{\sum_\ell S_\ell^2 R_\ell^2}{\sum_\ell S_\ell^2} \quad , \tag{3-29}$$

which can alternately be viewed as the variance-weighted average of indivi-dual R^2's. Again, in the dichotomous dependent variable case $S_1^2 = S_2^2$, $R_1^2 = R_2^2$ and $\mathcal{R}^2 = R_1^2$.

The multivariate Theta Statistic, Θ, generalizes the bivariate pre-

---

*A problem with modal prediction is the lack of sensitivity to statistical association of a type not altering the modal location which occurs often in distributions that are extremely unimodal at the univariate level. For a more detailed discussion see (Messenger and Mandell, 1972).

34

diction-to-the-mode concept to the multivariate level.* It is defined as the proportion correctly classed using a decision rule of predicting each individual as being in that dependent variable category having the <u>maximum</u> <u>forecast</u> value for that individual. For example, if an individual, whose dependent variable code Z=2, has a forecast { .15,.60,.25 } for codes 1, 2, and 3 respectively, then he would be correctly predicted as falling in category 2. This is expressed as

$$\Theta = \frac{\sum_{k} w_k \delta_{Z_k, \hat{Z}_k}}{\sum_{k} w_k} \quad . \tag{3-30}$$

In the absence of interaction this scheme is equivalent to the bivariate Θ_i calculated using a predictor variable comprised of all possible original predictor code permutations.** In the presence of interaction, the statistic is suboptimal since the least squares prediction criterion was used to determine the coefficients rather than a pure categorical prediction criterion. The classification scheme used by this statistic is felt, however, to be a useful alternative to that employed by multiple discriminant function (Nunnally, 1967) which has more restrictive assumptions. (See Chapter 4).

To aid in the interpretation of the Θ statistic, MNA computes and prints the proportion of cases correctly classed within each of the G

---

*The multivariate Theta statistic can also be interpreted as a linear transformation of a multivariate extension of the previously discussed Lambda statistic. In particular, with no interaction a multivariate Lambda statistic, λ^*, computed on a pattern variable comprised of all predictor code permutations is given by

$$\lambda^* = \frac{\Theta - \theta.}{1 - \theta.}$$

**This is termed a "pattern variable", and is discussed in Chapters 5 and 6.

dependent variable codes, and a classification matrix giving the frequency

for all permutations of predicted and actual codes.   (See Appendix C.)

Chapter 4

RELATIONSHIPS TO OTHER STATISTICAL TECHNIQUES
FOR ANALYSING NOMINAL SCALES

4.0 Summary

The problem of multivariate analysis of nominally scaled dependent
variables is not new, and various approaches have been suggested for doing
such analyses. It is the purpose of this chapter to identify some of the
more prominent and/or recent of such suggestions, and to indicate MNA's
relation to each.

The techniques to be discussed include the following: Multiple Classification Analysis (MCA), dummy variable multiple regression, Multiple
Discriminant Function Analysis (MDF) in both its usual and dummy variable
forms, Regression Estimation of Event Probabilities (REEP), Multiple Discriminant Analysis (MDA), the analysis of multidimensional contingency
tables (partition of chi-square and proposals by Goodman), the pattern-
probability model, and various recent suggestions by Coleman, Theil,
and Morgan and Messenger.

4.1 Characteristics of MNA

Before comparing MNA to other techniques for multivariate analysis
of nominal scales, it will be useful to mention some general aspects which
characterize MNA.

MNA is designed to be relevant for "theoretically oriented" or
"conceptually oriented" analysis. Some analyses have as their purpose
the development of a rather full understanding of how each predictor variable "works"--i.e. what is the effect of membership in each category of
a predictor, and how a given predictor as a whole relates to the dependent
variable when the predictor variable is considered alone, jointly with

others, and as supplemental to others. Other analyses are more "practically" oriented and concerned principally with arriving at an optimal prediction of the dependent variable. Unlike most of the other techniques to be discussed, MNA is "conceptually oriented." While this does not impair its ability to produce good predictions using the predictor variables specified (within the confines of an additive model), it means that the set of independent variables, and their categories, actually used in the prediction equation is determined by the analyst rather than by some selection mechanism built into the technique.

A second characteristic of MNA is its ability to analyse relatively large numbers of predictors (up to 50) with moderate sized data sets (300 to 6,000 cases). To be able to handle many predictors with data sets of this size requires the imposition of one or more restricting assumptions. MNA uses the additivity assumption.

A third characteristic of MNA is its focus on mangitudes of relationship rather than the statistical significance of those relationships. It is the authors' experience that with data sets in the size range for which MNA is intended, almost any relationship which is large enough to be important is likely to be statistically significant.

And fourthly, we were concerned to have the MNA output easily interpretable. This is again relevant to the basic conceptual orientation of the technique. We believe that what MNA has to say about a given batch of data is more easily grasped than are results which derive from many of the other techniques discussed in this chapter.

4.2 MNA, MCA, and Dummy Variable Multiple Regression

Multiple classification Analysis (MCA) (Andrews, et al. 1967) and dummy variable multiple regression (Suits, 1957) are essentially similar statistical techniques with some modest (though important) differences in the form in which results are shown. Both are intended to be used on similar kinds of data and they both have similar assumptions and restrictions. Thus, it is convenient to group them together when relating them to MNA.

The relationship of MNA to MCA and dummy variable multiple regression is extremely close. MNA is actually a series of MCA's run in parallel, with some additional summary statistics derived from MCA-generated results. Instead of analysing one intervally scaled dependent variable (as does MCA), the nominally scaled dependent variable is converted to several 0-1 dummy variables—one dummy variable for each category of the original nominal scale, and parallel MCA's are run using each of the dummy variables in turn as a dependent variable. It is appropriate to use MCA on these dependent dummy variables because each does, in fact, meet the requirement of an interval scale: all intervals are equal (in this case there happens to be just one interval). The dependent dummy variables each indicate simply whether a particular case is—or is not—in a given category of the original nominal scale. Thus all of the strengths and weaknesses of MCA are also applicable to MNA.

MNA adds, however, several features not present in MCA. Most important is the behavior of the coefficients and the forecast probabilities across the categories of the dependent variable. As noted in previous chapters, the coefficients associated with any one predictor variable category sum to zero across the categories of the dependent variable, and the forecast probabilities sum to 1.0 across these categories. These are not specially imposed constraints, but simply a property of running a complete set of parallel MCA's with identical predictors. They give MNA, when taken as a whole, a coherence and interpretability which is not attributable to any one of its MCA constituent parts considered alone.[*]

In addition to the results obtained from the parallel runs of MCA, MNA includes a number of additional summary statistics: bivariate and multivariate Thetas, generalized eta's, generalized multiple correlations, and a classification matrix representing categorical prediction agreement and disagreement.

4.3 MNA and Multiple Discrimination Function Analysis (MDF)

The most widely known multivariate technique for analysing a nominally scaled dependent variable is multiple discriminate function analysis (MDF).[**] In its usual form, MDF requires intervally scaled predictor variables whose relationships among themselves are linear. These restrictions on the independent variables are severe and eliminate MDF in its usual form as a useful technique for analysing many kinds of data. As noted in Chapter 2, MNA is not limited by these restrictions. MNA can accept data where independent variables are at any level of measurement and where relationships among predictor variables have any form whatsoever. In fact, much of the original stimulus to develop MNA came from a desire to do multivariate analysis of nominal scales on data for which MDF was known to be inappropriate.

MDF, however, can be used with dummy variable predictors, and in this form is not bound by the above restrictions. Dummy variable multiple

---

[*] Section 3.3 describes the transformation which is applied to the dummy variable multiple regression coefficients which converts them to MCA-like coefficients and gives them their ease of interpretability.

[**] This technique is based on work by Fisher (1936) and is described in many modern texts including Overall and Klett (1972).

discriminant function analysis and MNA have, in fact, many features in common. Both are appropriate for the same kinds of data, both assume an additive model (i.e. that interaction is not present), and from the predicting standpoint both should produce comparable results.[*]

The means by which MNA and dummy variable MDF arrive at their predictions for each case, however, are somewhat different. Essentially, MNA derives the "effect" of membership in each category of each predictor variable, allowing for interrelationships among the predictors, and simply adds up the effects to arrive at a forecast for each individual. From this a prediction of category membership can be made. The various figures which result--indications of the effects of particular categories and coefficients showing univariate and multivariate strength-of-relationship between the dependent variable and the independent variables--are easy to interpret and may have substantial interest in themselves.

In contrast, MDF--in allowing for the interrelationships among the predictors--derives a new set of predictors (the "discriminant functions") which are then used to make the predictions. These have the statistical property of being orthogonal to one another, but are arbitrary from the conceptual standpoint and may in fact be uninterpretable. Thus while MNA and dummy variable MDF seem comparable if the goal is predicting

---

[*] To date we know of only one empirical comparison between MNA and dummy variable MDF (O'Malley, 1972). This comparison involved three independent variables, each with three categories, a three-category dependent variable, and 1371 cases. MNA correctly predicted the dependent variable category in 58.3% of the cases, while MDF correctly predicted 57.5% of the cases. Although the predictive power of the two techniques was about the same, the patterns of errors (wrong predictions) differed substantially.

the category of a dependent variable to which a given case should be assigned, we believe that MNA is the more useful technique when the analytic goals include a search for understanding about the particular effects of the independent variables taken singly and jointly.

4.4 MNA and Regression Estimation of Event Probabilities (REEP)

A statistical analysis technique which has substantial similarities to MNA has been developed by Miller (1964) and is embodied in a series of computer programs known as REEP (Regression Estimation of Event Probabilities). REEP was stimulated by meteorologists' needs for predicting weather conditions--which are treated as the categorical dependent variable.

Like MNA, REEP converts a nominally scaled dependent variable to a series of dummy variables, and runs a series of parallel multiple regressions using each of the resulting dummy variables as dependent variables. Also like MNA, REEP first converts the predictor variables to dummy variables. MNA and REEP are intended for the same kinds of data, process it using basically similar methods, and are similarly restricted by the assumption of additivity.

An important difference in orientation, however, distinguishes REEP and MNA. REEP's primary purpose is to produce forecasts, with less emphasis on helping the analyst get a full understanding of the effects of each of his predictor variables. While MNA also produces forecasts, it does so in a way which assists the conceptually oriented analyst in seeing how each of his variables operates vis-a-vis the dependent variable.

This difference in orientation is reflected in REEP's incorporation of an automatic "screening" of predictors. REEP will accept predictor variables which, after dummyizing, may result in up to 500 dummy variables. The screening process determines which of these dummy predictors is most related to the dependent variable, then which one dummy predictor adds the most explanatory power to the variable already selected, which one then adds most to the two already selected, and so on. Up to 36 dummy predictors may be selected by this screening process, and the parallel regressions are then carried out using the selected predictors. It is quite possible that REEP's screening process will select some dummy variables from one original predictor variable but not all--e.g. the dummy variable indicating whether a respondent came from the South might be selected, but other region categories might not be. Since the analyst has no control over which dummy predictors will be selected, it may be difficult for him to get a clear understanding of how particular variables he chose for their conceptual importance relate to the dependent variable.

In contrast, MNA accepts up to 100 dummy predictors and conducts the analysis using all that were originally submitted with no procedure for intermediate screening.

4.5 MNA and Multiple Discriminant Analysis (MDA)

In addition to REEP (discussed above) Miller and his colleagues have produced another approach for the multivariate analysis of a nominally scaled dependent variable, known as Multiple Discriminant Analysis (MDA) (Miller, 1962). MDA is similar to REEP in its strong emphasis on forecasting and is not particularly useful for the more conceptually oriented analyst.

Like REEP, MDA screens up to 500 dummy variable predictors, sequentially selects up to 36 that are most useful for forecasting, and then develops the forecasts. Instead of using additive regression models for forecasting, however, MDA examines the distribution of the dependent variable in each unique combination of the predictor variables. Up to 2500 distinct combinations of predictors can be examined. (If the selected predictors result in the actual occurrence of more than 2500 patterns, predictors are dropped--sequentially in reverse order from their selection--until less than 2500 patterns occur.) The forecast is based on the distribution of the dependent variable among the cases which show similar patterns on the predictors, or which are located "near" those cases in the multidimensional space defined by the predictors.[*]

The individual examination of all possible combinations of predictor variables is an elegant way of avoiding the restrictions of an additive model but is achieved at other costs. The costs include restricting the analysis to a relatively modest number of dummy predictor variables, the use of very large data sets, the "fuzzing up" of the predictors by the necessity of basing a prediction on cases from more than just a single combination of predictor variables, and the likelihood that measurement or sampling errors present in the small groups with unique patterns will produce idiosyncratic results.

[*] MDA includes an intermediate dimensionalizing of the dummy variable predictors which were selected by the screening test. In this respect it somewhat resembles the multiple discriminant function analysis discussed previously.

4.6 MNA and the Analysis of Multidimensional Contingency Tables

Still another approach for the multivariate analysis of nominally scaled dependent variables is through the analysis of multidimensional contingency tables. In early stages of development, this consisted of partitioning the statistic chi-square. This technique was pioneered by Irwin (1949) and Lancaster (1949, 1951), described in the psychological literature by Sutcliffe (1957), and has been the topic of a series of recent articles by Goodman (1969, 1970, 1971, 1972a,b) and by Grizzle Starmer and Koch (1969). The recent work has moved away from the partitioning of chi-square and proposes alternative approaches for handling multidimensional contingency tables.

The techniques provide powerful ways to assess whether there are significant main effects or interactions attributable to a limited number of independent variables in relation to a nominally scaled dependent variable. Rarely are they applied to more than a few independent variables at a time, and rarely do the independent variables have more than three or four categories. As many as 16 categories across the independent variables would constitute a large problem. One reason for these limits is that the techniques use the data in each "cell" of the multidimensional contingency table--i.e., in each particular combination of the independent variables--and thus require a very large number of cases if applied to more variables and/or more categories.

MNA's ability to handle many more independent variables (up to 100 categories across the independent variables) with moderate numbers of cases represents one major difference between MNA and methods referenced above. MNA gains this ability by the use of an additive model, a restriction of MNA which is not present in the partition of chi-square and the more recent proposals cited. Of course, if the additivity assumption is introduced when analysing multidimensional contingency tables, this approach can handle the same number of predictors as MNA can, but then loses its power to handle interactions.

Another difference between MNA and the partition of chi-square is that MNA focuses on the <u>strength of association</u> between independent and dependent variables and magnitude of "effects" of particular categories of the independent variables, whereas the partition of chi-square focuses on statistical significance--i.e., the probability that an observed relationship could have occurred by chance alone.[*] Goodman's recent writings however, include proposals of ways to measure the magnitudes of effects in multidimensional contingency tables.

4.7 MNA and the Pattern-Probability Model

Overall and Klett (1972, Chapter 16) propose an interesting way to predict which category of a nominally scaled dependent variable a case will be located in by use of what they call the "pattern-probability model." The model requires the assumption that the predictor variables are statistically independent of one another (i.e., unassociated) <u>within</u> groups of the dependent variable (though they may be interrelated when cases from several different groups are combined). If the data meet this assumption, Overall and Klett show that one can derive useful

---

[*]Analysts differ on whether one should focus primarily on magnitudes or significance. As noted earlier, we find focussing on magnitudes to be more productive with the size data sets for which MNA is intended. Morrison and Henkel (1970) provide a compilation of articles relevant to the role of statistical significance in data analysis.

forecasts by computing the probability of occurrence of a particular pattern of predictor variable characteristics in each category of the dependent variable. What is distinctive about their approach is that the assumption of within-groups independence permits computation of the required probabilities without the necessity of having a huge data set even when using large numbers of predictor variables. It would appear that the assumption of within-groups independence provides for this approach a "leverage" which is analogous to the leverage gained in MNA by its assumption of additivity. Of course, the usefulness of both approaches is restricted by the validity of its particular assumption.

In addition to the difference in fundamental assumptions, several other differences between MNA and the pattern-probability model can be mentioned. Like REEP and MDA, discussed earlier, but unlike MNA, the pattern-probability approach is primarily intended for generating predictions. It does not produce measures of the effect of membership in each category of each predictor variable, nor measures of the strength of association of each predictor variable to the dependent variable which would be of central concern in more theoretically oriented analysis.

4.8 Suggestions Involving Transformations of Proportions

Several recent articles on multidimensional contingency tables have suggested that in analysing the proportion of cases which fall into a given category of a nominally scaled dependent variable one should first apply a transformation to the proportions (e.g., Theil 1970, Walker and Duncan, 1967). Several different transformations appear in the literature. Logits and probits seem the most common, and of these two the logistic transformation seems to have largely replaced the probit. The logistic function has the effect of transforming a probability value, which is limited to the range 0 to 1, to a number which is unbounded, from $-\infty$ to $+\infty$. Thus, one is not faced with the possibility of encountering a "probability" which is outside its logical range.

Several considerations led us _not_ to incorporate a logistic transformation in MNA. First, the use of logits requires the prior calculation of a proportion, which itself requires the assignment of individuals to one of several groups. MNA was intended to operate directly on the scores of individual cases, rather than on some prior multidimensional contingency table, and it was not clear to us how logits would be determined, given the large number of independent variables we wished MNA to accept and our desire to keep the micro character of the data.

Secondly, we believed that logits were most useful at the ends of the proportion range--i.e., when proportions were close to 0 or 1--and we believed that the problems encountered at the ends of the range could be minimized by avoiding attempts to analyse dependent variables including a category with an extremely large or small proportion of cases falling into it. (Hence, our advice that each category of the dependent variable include at least 10% of the cases.)

Thirdly, we knew that MNA would not produce forecasts including probabilities outside the 0-1 range for any actual case so long as the data met the additivity assumptions of MNA. Thus the "artificial" avoidance of such values by use of a transformation seemed unnecessary to us.

Fourthly, we wished to keep MNA and its output as conceptually

straightforward as possible. The incorporation of a transformation un-
familiar to many potential users seemed undesirable unless it were shown
to be absolutely necessary.

It is possible that someone will show that prior calculation of
proportions, and subsequent analysis of them in a transformed form, will
provide a multivariate analysis approach superior to MNA. As of this
writing, it is not clear to us how this might be done on the kinds of
data for which MNA is intended, nor that there would be advantages in
attempting to do so.

4.9 Suggestions by J.S. Coleman

J.S. Coleman (1964, 1970) is among those who have been concerned
with the analysis of "attribute data" (to use his term). He builds an
analysis model on the assumption of there being "effects" exerted by one
or more independent variables and also "random shocks" which "move"
objects from one category to another on the dependent variable and which
ultimately determine the stable proportions which are actually observed
in the data.

Most of his writing on this topic assumes that the dependent variable has just two categories, and there is little treatment of the more
complex situations which MNA is designed to handle. However, like MNA,
his basic approach assumes an additive model.

Furthermore, some of his statistics approach those incorporated in
MNA. Coleman writes:

> Since the present approach results in a linear model,
> it would be surprising if it were not related to other
> statistical procedures based on a linear model. In
> fact, it can be shown that its effect parameters... are

> identical to regression coefficients calculated using
> multiple regression analysis, when the independent vari-
> ables take on only the value 0 or 1 ("dummy variables"),
> and the dependent variable is a proportion. ...The weight-
> ed estimates are identical to regression coefficients based
> on regression analysis carried out with individual observa-
> tions. (Coleman, 1970, p. 232.)

To our knowledge, Coleman has not incorporated his proposals into
an integrated computer program for the actual analysis of large data sets
with polychotomous dependent variables and many independent variables.
Presumably this could be done, and it would be interesting to compare the
results with those produced by MNA.

4.10 MNA and THAID

Morgan and Messenger (1973) have been developing an analysis technique--
known as THAID--which, while totally different from MNA in operation, is intended for similar kinds of data. THAID provides a series of sequential binary
divisions of the cases (each division being defined in terms of the categories
of one independent variable) such that the distribution of cases on the dependent variable is "maximally different" in the resulting groups.[*] Thus, each
of the sequential splits is chosen to provide the maximum possible "explanation" of the dependent variable within the group of cases being considered.
THAID is intended to be used on nominally scaled dependent variables in much
the same way that AID (Sonquist, et al., 1971) is used on intervally scaled
dependent variables.

THAID provides a useful complement to MNA. THAID does not assume an
additive model, and can be used to identify some of the non-additivities
which would make difficulties for MNA. THAID's primary orientation is to

---

[*]What is maximized is the frequency weighted sum of the absolute differences
between the original proportions and those in the groups resulting from the
divisions.

search out an optimal model for explaining differences between cases with respect to the dependent variable -- i.e., to identify the underlying "structure" in the data. In contrast, MNA assumes one particular structure (that of an additive model) and provides estimates of its parameters.

Sonquist (1970) has written about the complementarity of the two analysi strategies AID and MCA for intervally scaled dependent variables. We believe that much of what he has said can also be applied to the THAID-MNA pair when the dependent variable is nominally scaled.

Chapter 5

DESCRIPTION AND USAGE OF THE MNA COMPUTER PROGRAM

5.0 Summary

The first objective of this chapter is to give the user an overview of
the MNA program, referencing more detailed information such as setup instructions, flow charts, etc. which appear in the appendices. Second, the re-
strictions--program, numerical and theoretical--are discussed. Third, a
basic user strategy is outlined. And fourth, use of the option to compute
residuals is described. Notation in this chapter is the same as that defined in Basic Terms of Chapter 3.

5.1 Program Design

The MNA program is written in single precision FORTRAN G and uses standard OSIRIS subroutines (University of Michigan, 1973b) written in IBM
Basic Assembly Language and FORTRAN G.* It presently utilizes most available core in a 104K MFT/HASP partition in an IBM 360/40, up to two tape
drives, two disk datasets, card reader and printer. The program is over-
layed and has a length of $13,950_{16}$ bytes. Non-overlayed length is approximately $1C,000_{16}$ bytes.

The program normally reads input data from tape in character mode and
writes an integer mode converted subset to be used in any subunit of analysis (analysis packet), as specified by the list containing variables to be
analyzed (variable list) and optional case subsetting card (global filter),
into a first disk dataset. The data may have been recoded during the tape

---

*OSIRIS is a set of programs and supporting subroutines developed
and distributed by the Institute for Social Research, University
of Michigan. Readers should refer to the referenced documents for
detailed explanation of terms.

read and conversion process using the multivariate recode facility (University of Michigan, 1973a, Appendix B).

Next, analysis packets are processed by a further subsetting of data to (optionally) exclude analysis packet filtered, outside of range, or cases matching missing data codes (missing data cases) and written onto a second disk dataset. During this process, generation of the sums of squares cross products matrix occurs.

The program then calls a series of subroutines that solve the normal equations by first inverting the sums of squares-cross products matrix and then reapplying that inverse, common to all G regression equations, to each of the G regressions. The need to perform an accurate matrix inversion on matrices up to 100 X 100 constrained by a 104K memory led to a considerable search of matrix inversion routines. The choice was Gaussian elimination with partial pivoting and row equilibration (scaling) written in FORTRAN G (Forsythe and Moler, 1967). This routine does not assume symmetry and has an iterative improvement procedure allowing the matrix to be stored in single precision (32 bit word length) with critical calculations done in double precision (64 bit word length), such that the accuracy approaches that of most non-iterative double precision routines. The routine prints a message indicating the minimum number of significant digits in the solution vectors. As a test, successively larger Hilbert matrices (ibid, p. 80) were input with the routine breaking down and unable to invert the 9th order Hilbert matrix. In practice no inversion failures have occurred to date and matrices up to 90 X 90 have been successfully inverted in tolerable times (3 minutes for 90 x 90).

The final step involves a read from the second disk dataset to compute and print the summary statistics. Other analysis packets are processed in the same way until the end of the card input stream is reached, when a normal termination message is printed.

Detailed set up instructions are listed in Appendix A, the macro flow chart is shown in Appendix B, and sample output is given in Appendix C. Appendix D provides information on obtaining and adapting the program.

5.2 Restrictions

Restrictions for using the program can be categorized into program restrictions, numerical restrictions that force the use of data with certain requirements to guarantee accuracy of the calculations, and finally theoretical restrictions based on assumptions in the least squares regression technique.

5.2a Program restrictions.

There are four major program restrictions built into MNA. These are:

A. Number of non-empty dependent variable codes $\leq$ 10. (Each analysis packet).

B. Number of non-empty independent variable* codes $\leq$ 20 per variable.

C. Total number of independent variable codes $\leq$ 100. (Each analysis packet).

D. Number of variables to be analyzed $\leq$ 100. (Total for all analysis packets).

5.2b Numerical restrictions.

There are two numerical restrictions:

E. The weighted number of cases $\leq$ 1,000,000.

F. Independent variables having one or more "perfectly overlapped categories" (See Tables 5.1, 5.2) must be modified.

Restriction E is to insure the numerical accuracy of the various cal-

---

*"Independent variable" is interchangeable with "predictor variable".

culation routines within MNA. It is beyond the scope of this monograph to include the details of the mathematics required to develop the reasons for this constraint. Suffice it to say that the problem is complex with several unpleasant properties of the exact solution such as dependence on data entry order and distribution of magnitudes of the variables, etc. (Ibid, pp. 87-108). This constraint was arrived at in a conservative way giving a generally applicable while still tolerable restriction.

The inversion routine has a parameter allowing the minimum number of significant digits in the solution vector to be specified. If this minimum level cannot be attained, the error message ****WARNING MATRIX DOES NOT CONVERGE**** is printed indicating probable numerical problems. This level is set sufficiently high so that if the message does not appear, the only source of numerical errors is in the sums of squares-cross products matrix. Because all variables in the sums of squares-cross products matrix are dummyized specifically to be 0 or 1, this matrix is relatively invulnerable to numerical problems. Again, users need only be concerned if weight sums for their data sets exceed one million (Restriction E).

Restriction F is present because dummy variable predictors that have linear dependencies result in a singular sums of squares-cross products matrix. The program will print the message SINGULAR MATRIX IN DECOMPOSE. ZERO DIVIDE IN SOLVE and abnormally terminate. The simplest and by far the most common cause of this is the perfect positive, 1.0, correlation of two dummy predictors, termed "perfectly overlapped categories." This is illustrated in the bivariate table of predictors X_i and X_j in Table 5.1.

Table 5.1

Perfectly Overlapped Categories (r=1.0)

X_j

	1	2	3	9	
1	5	3	2	0	10
2	20	15	20	0	55
3	12	8	10	0	30
9	0	0	0	5	5
	37	26	32	5	100

X_i labels the rows.

This table shows a hypothetical joint frequency distribution for a sample of 100 cases. The two dummy variables associated with codes 9 of both X_i and X_j are simultaneously either 0 or 1 and hence have a sample correlation of 1.0. The code 9 was chosen for illustration since missing data is frequently assigned this code and it is fairly easy to obtain perfectly overlapped categories in this manner unless one exercises caution. It is also possible, but unlikely, to obtain sample correlations of -1.0. Table 5.2 illustrates this for codes 1 and 3 respectively of predictors X_i and X_j.

Table 5.2

Perfectly Overlapped Categories (r=-1.0)

$$X_j$$

		1	2	3		
	1	20	10	0		30
	2	0	0	10		10
X_i	3	0	0	10		10
		20	10	20		50

It is good precautionary practice to examine all bivariate predictor tables in which suspected overlapped categories exist before running MNA. If one encounters a singular matrix a first step is also to examine the bivariate tables (not part of the MNA output).

5.2c Theoretical restrictions. The theory underlying least squares regression places no restriction on the distribution of the dependent variable in order to obtain the best linear unbiased estimates (B.L.U.E.) of the coefficients (Bohrnstedt and Carter, 1971, p. 122). In order to do hypotheses testing, however, it is formally required that the dependent variable be normally distributed. However, multiple regression techniques have been found to be robust with respect to violations of this requirement (Ibid., p. 123), and it is suggested that no category of the original dependent variable contain less than 10% of the cases. If this restriction is not met the variance of the regression coefficients becomes prohibitively large.

Secondly, the sample size should be sufficient to support the estimation of the typically large number of coefficients in the model. Excessively large numbers of dummy predictor variables, relative to the sample size, gives inaccurate coefficients and summary statistics in terms of describing the population from which the sample was drawn. A rule of thumb is not to have less than 10 times as many cases as number of dummy predictor variables, C. For example, with 300 cases one should try to keep the number of dummy predictor variables to 30 or fewer (e.g., not more than 10 predictors each of which has 3 categories, 6 predictors each with 5 categories, etc.). This restriction is felt to be a compromise allowing a reasonable model size while holding both the bias and variance of the coefficients and summary statistics to a tolerable level.

To illustrate the reasoning behind the rule consider the expression for the adjusted squared multiple correlation, $R_{\ell}^{2'}$, alternatively represented by

$$R_{\ell}^{2'} = 1-(1-R_{\ell}^{2}) \frac{N-1}{N-C+P-1} \qquad .$$

By defining $K = \frac{N-1}{C-P}$, we can re-write as

$$R_{\ell}^{2'} = 1 - (1-R_{\ell}^{2}) \frac{K}{K-1}$$

An <u>arbitrary</u> but useful limit on the tolerable decrement to R_{ℓ}^{2} has been set at 20%. Thus we can plot K versus R_{ℓ}^{2} to achieve this 20% decrement. The results of this appear in Figure 5.1.

Figure 5.1

Cases Relative to Predictor Categories

Needed for 20% R_ℓ^2 Shrinkage

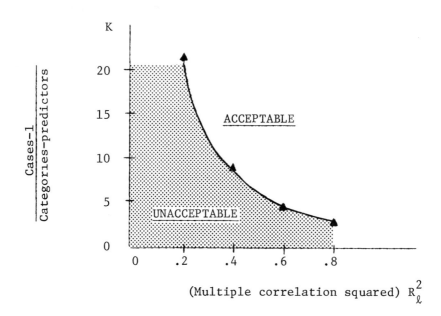

Users can determine acceptability before their analysis by estimating R_ℓ^2 and determining from the values of N, C, and P whether or not their analysis will be acceptable in terms of the stated criterion of a 20% R_ℓ^2 decrement. The 10:1 rule of thumb is based on typical R_ℓ^2's of .3-.4 and $\frac{C}{P}$ ratios in the neighborhood of 4-5.

Thus, the following two constraints are added by theoretical and sampling considerations.

G.   Each dependent variable code contains at least 10% of the data.

H.   There be at least 10 times as many cases as dummy independent variables.

5.2d   Assumption of additivity.   The theory underlying MNA also assumes that the prediction of the dependent variable is derived by simply adding together the separate effects of membership in a particular category of each independent variable. Thus we have the following restriction:

I.   The data are appropriately described by an additive model, i.e.,

no substantial interactions occur in the data.

A simple definition of interaction is that the relationship between

the dependent variable and a predictor depends on one or more other pre-

dictors. If we were to find, for example, that education had one effect

on people's orientation toward violence among whites, and a different ef-

fect among blacks, we would have identified an instance of interaction.*

The above example of interaction involves just two predictor variables

(education and race) and is sometimes referred to as "simple" or "first

order" interaction. Higher order forms of interaction, involving three

or more predictor variables are also possible.

From an analytic standpoint, Restriction I is fundamental. While

additive models have been shown to appropriately describe numerous real

phenomena, interaction effects are also widely present. When a relatively

simple interaction effect can be identified (two or three predictors and

the dependent variable), it is usually possible to combine certain vari-

ables into a new "pattern variable" (see definition below), which may

then be appropriately handled. However, if interactions are present in

the data, either because the analyst does not know they are there,

or because he has not constructed needed pattern variables, they will

result in misleading MNA results. The indicated effects of certain pre-

dictor categories will not reflect actual effects, predicted scores will

be erroneous, and statistics summarizing the strength of bivariate and/or

_____

*Other terms sometimes used to refer to interaction include
"moderator effects," "contingency effects," and "conditioning
effects."

multivariate relationships will indicate lower degrees of association than are actually present.

In most real data all forms of interaction exist to some degree. The problem for the MNA user is to determine whether substantial interactions are present which might degrade his results if not appropriately entered in the analysis. As is suggested in Chapter 6, the identification of interactions involving nominally scaled dependent variables is an analytic area needing further development. Here we will simply indicate four possible approaches. Once an instance of interaction has been identified, it can be further examined by an approach we will describe to see whether the effect is substantial.

One source for identifying possible interactions is the substantive theory underlying the research. Based on what the analyst knows or hypothesizes about his variables, he may be able to predict among which variables interactions will occur.

A second approach is to use another analytic technique to help identify occurrences of interaction on an empirical basis. Chapter 6 identifies several which may help in doing so—THAID, MDA, and analyses of multidimensional contingency tables. (Of these, THAID is most likely to be appropriate for the same kinds of data as would be submitted to MNA.)

A third approach is to examine an initial MNA run to see whether any forecast values fall outside the range 0-1. If any values fall outside this range, some interaction must be present. However, the fact that all values fall within the range does not prove that interaction is absent.

Still a fourth approach is the tedious and expensive testing of all possible pairs, triplets, etc. of predictor variables using the pattern variable technique we describe next. While this approach may be useful when the number of predictor variables is very small, its feasibility declines rapidly as the number of predictor variables increases.

The following "pattern variable" strategy has proved useful in determining the magnitude of an unknown interaction. First, an MNA run is made using the predictors in original form. The resulting R^2 indicates the degree of predictability when no interaction effects are reflected in the analysis. Next a "pattern variable" comprised of all predictor code combinations is formed. Table 5.3 indicates how a pattern variable, X', would be formed by combining two predictor variables, each of which has three categories.

Table 5.3

Example of Forming a Pattern Variable
from Two Predictor Variables

Predictor X_1	Predictor X_2	Pattern Variable X'
1	1	1
1	2	2
1	3	3
2	1	4
2	2	5
2	3	6
3	1	7
3	2	8
3	3	9

(Note that the number of categories in the pattern variable is given by the product of the number of categories in each predictor variable. If there are many predictor variables being combined, and/or if there are many categories in each predictor variable, this product gets large and may itself become unanalysable. For example, five predictors with three codes each would result in a pattern variable with 243 categories.)

Once the pattern variable is formed, it is entered in an MNA analysis. Since its categories now include all the additive plus all the interactive effects of its constituent variables, any gain in the resulting R^2 over what was obtained when the variables were entered separately is a reflection of the interaction effects. If the gain is small enough to be judged inconsequential, the analyst can use MNA knowing interactions among the variables tested do not pose problems in his data. If the gain in the generalized multiple R-squared is judged to be worth taking notice of, the analyst may still be able to use MNA by introducing appropriate interacting variables into MNA runs in their combined, pattern variable, form.*

An example of a set of analyses using the pattern variable approach to test for and identify interaction is indicated in Table 5.4.

---

*This latter statement is phrased rather tentatively because the authors have encountered instances in which the pattern of interactions was so complex that no simple combining of two or three variables was possible which might then be analysed together with several other separate variables. The pattern variable approach is most useful when a small number of predictor variables interact with each other but not with any other predictor variables.

Table 5.4

Series of Runs Illustrating the Use of Pattern
Variables to Identify Interaction Effects

Note: "1X2" indicates a pattern variable
formed by combining predictors 1 and 2.

Run	Predictors	Resulting R^2
1	1,2	.20
2	1X2	.20
3	1,3	.15
4	1X3	.17
5	2,3	.23
6	2X3	.34
7	1, 2X3	.44
8	1X2X3	.45
9	1,2,3	.30

The fact that R^2 does not change between Runs 1 and 2 indicates predictors 1 and 2 do not interact.* The small change between Runs 3 and 4 indicates little interaction involving predictors 1 and 3. The difference in R^2 between Runs 5 and 6, however, indicates the need to consider the interaction involving predictors 2 and 3. Run 7 provides an R^2 incorporating all important first order interactions. If the number of cases permits, one could go on to Run 8 and test for the only other interaction which might be present in these data (a single second order interaction involving all three predictors). In this example, the gain in R^2 over Run 7 is minimal and one would conclude that second order interaction presented little problem. (If one had made Runs 1-8, there would be little point in making Run 9, though this is a logical possibility.)

---

*Although we have designated the results of Run 2 (and also of Runs 4, 6 and 8) as R^2's, and while they can be obtained from MNA as R^2's, they are identical with generalized eta squares obtainable from a conventional one-way analysis of variance (or from MNA).

The sequence of runs described above is intended to illustrate the general principles of the pattern variable approach. In practice, if one were using only three predictors, a more efficient series of runs would be those indicated in Table 5.5. These would lead the analyst to the same conclusions using fewer runs.

Table 5.5

Series of Runs to Identify Interactions
Among Three Predictors

Note: "1X2" indicates a pattern variable
formed by combining predictors 1 and 2.

Run	Predictors	Resulting R^2
1	1,2,3	.30
2	1X2X3	.45
3	1X2,3	.30
4	1X3,2	.31
5	1,2X3	.44

Run 1 shows how well one can do without taking into account any interactions. Since Run 2 produces an R^2 which is substantially higher than Run 1, there is evidence that some interaction is present. Runs 3 and 4 produce little gain over Run 1, indicating that the interaction does not involve predictors 1 and 2, or 1 and 3. Since Run 5 results in an R^2 almost as high as Run 2, one knows that almost all the interaction is a first order interaction involving predictors 2 and 3.

One problem with the pattern variable approach sketched above is that the generalized multiple R^2's on which the procedure depends are biased statistics. As described in Chapter 6, they are somewhat "too high," reflecting some capitalization on chance in fitting the statistical model. A more appropriate comparison would involve minimum variance unbiased estimators. As yet, the authors have been unable to identify an appropriate way to make the necessary adjustment. So long as Restriction H is adhered to, the amount of the adjustment would be rather small. In the absence of a better measure of interaction, the authors believe the above approach can be usefully applied.

5.3 User Strategy

Users of MNA should follow the series of steps outlined below when running the program. Doing this will uncover potentially expensive, time consuming problems in as early a stage as possible while indicating either the inapplicability of the program or a solution to the problem.

The first step is a careful examination of the variables to see if they are of the appropriate scale for MNA. The dependent variable should be nominal and the predictors categorical (either nominal or ordinal). If the dependent variable is intervally scaled, MNA is clearly not appropriate. If it is ordinally scaled, one has two alternatives. Either the variable can be assumed interval (Labovitz, 1970) and a technique such as MCA used, or the user can forego ordinal properties by using the variable in MNA anyway. The first method is more appropriate when the user wants his analysis to reflect "closeness" of prediction while the second is more appropriate if he is content to settle for a wrong or right categorical prediction criterion.

Given data appropriately scaled for MNA, the user should next examine the variable distributions against the MNA restrictions on code values, frequencies and proportions given in the previous section entitled Restrictions. Although the restrictions do not limit the frequencies in the predictor variables, it is good practice to check these and ensure that no predictor variable has categories with very small N's.

If predictors or dependent variables have inappropriate distributions one can often solve the problem by either (a) eliminating the low-frequency categories by filtering them out (of course this may destroy the representativeness of the sample) or (b) collapsing two or more small categories into a single larger one.

Given that the first two steps are successful, the user should check known interactions that violate the basic additivity assumption of MNA. If serious interactions are apparent a priori they should be dealt with by construction of an appropriate pattern variable or possibly by carrying out the analysis within separate subgroups.

Finally, MNA runs should be made, checked for interaction, and appropriate changes incorporated in a final run.

To summarize, the proper sequence of user steps is as follows:

A. Determine if variables have appropriate scales.

B. Check MNA constraints.

C. Check for and eliminate large a priori interactions.

D. Run program.

E. Examine output for interaction making changes and re-running if necessary.*

5.4 Residuals**

Residuals are optionally computed, printed and written according to the local parameter keywords RESIDUALS/NORESIDUALS, IDVAR=n and SUPPRESS/NO-

---

*The two basic tools the user has to work with are (a) an examination of forecast values falling outside the range 0-1, and (b) the comparison of the MNA R^2's using pattern variables. Both are described above in section 5.2d.

**For completeness, a brief discussion is given here on the subject of MNA residuals which are at present purely experimental. See also Chapter 6.

SUPPRESS (See Appendix A - Local Parameter Card). The definition of residuals in MNA is a set of G differences between the G dummy dependent variables and their corresponding forecast values, one set for each original
case. For example, in Table 5.6 the original data consists of the 3 cases
of ID variable values and the dependent variable values, Z. MNA generates
the dummy variables Y_1, Y_2, Y_3; computes the forecast $\hat{Y}_1$, $\hat{Y}_2$, $\hat{Y}_3$ and the
residuals Re_1, Re_2, Re_3.

Table 5.6

Example of Computation of Residuals

Original data		Dummyized dependent vars.			Forecast			Residuals		
ID	Z	Y_1	Y_2	Y_3	$\hat{Y}_1$	$\hat{Y}_2$	$\hat{Y}_3$	Re_1	Re_2	Re_3
1	3	0	0	1	.2	.0	.8	−.2	.0	.2
2	1	1	0	0	.7	.3	0	.3	−.3	0
3	2	0	1	0	.1	.7	.2	−.1	.3	−.2

Output format for residuals is the ID variable number followed by the
G residuals in corresponding dependent variable code order. For example for
case #1 in Table 5.5, the record containing the residuals would have {1,-.2,
.0,.2 }.

The motivation behind use of residuals is an attempt to generalize to
nominal scale dependent variables the successful residuals methodology applied to least squares regression in general and the MCA dummy variable
program in particular.

Chapter 6

UNSOLVED PROBLEMS

6.0 Summary

This chapter presents short discussions about five unsolved problems associated with the use of MNA. Some are not problems unique to MNA but represent problems of multivariate data analysis more generally. As implied by the chapter title, we have no final solutions to offer here. Rather, the purpose is to indicate the direction of our thinking and the thrust of some methodological investigations planned for the future, and to invite suggestions for possible solutions.

The problems discussed are: (1) modes of searching for interactions, (which, if present, violate one of the fundamental assumptions of MNA); (2) adjusting summary statistics for degrees of freedom--i.e., for the extent to which they capitalize on chance in fitting the model; (3) testing for statistical significance; (4) using residual scores from MNA; and (5) determining the importance of predictor variables.

6.1 Modes of Searching for Interaction

MNA, like a number of the other analysis techniques discussed in Chapter 4, assumes that an additive model adequately represents the data being analyzed. As noted previously in this monograph, the results from MNA will be misleading if the data contain substantial interactions which are not allowed for in preparing the variables for MNA.

It seems unlikely to us that MNA can be made to handle complex interactions involving many variables simultaneously. However, if relatively simple interactions (e.g. involving just two or three predictor variables and the dependent variable) are identified in advance, it is usually feasible to combine the variables involved in a way which will let MNA handle them within the context of an additive model. (The technique is discussed in Chapter 5.)

Thus one would like answers to the following questions: (1) Are any complex interactions with substantial effects present? (2) Are any simple interactions with substantial effects present? (3) If simple interactions with substantial effects are present, what predictor variables are involved?

A variety of ways for identifying interactions have been suggested. As noted in Chapter 4, work in the tradition of analyzing multivariate contingency tables, particularly recent work by Goodman (1972a, 1972b), provides an elegant approach. While we suspect this approach may prove infeasible if the number of predictors is a half dozen or more and the number of cases is in the range found in typical surveys, the approach surely deserves further investigation. MDA (Miller, 1962) is an alternative approa

Chapter 4 also discusses the THAID approach (Morgan and Messenger, 1973). This approach is specifically designed for use on typical survey data, and promises to be useful for identifying major interactions involving predictors which show above average relationships to the dependent variables. However the form of the THAID output sometimes makes identification of these interactions difficult, particularly for interactions involving predictors on which THAID does not "split."

If a way could be found to know that complex interactions were not present, the task of testing for simple interactions involving particular variables is probably solvable. Chapter 5 lists two approaches--one involving examination of forecast values and the other an analysis of pattern variables. Unfortunately, neither is simple to carry out with present computer software if the number of predictors is large. The area is ripe for further investigation.

6.2 "Adjustment" for Degrees of Freedom (Unbiased Estimators)

MNA computes bivariate and generalized eta-squares, $\eta_{i\ell}^2$ and η_i^2, and multivariate and generalized R-squares, R_ℓ^2 and $\mathcal{R}^2$, all of which are known to be biased estimators. While correctly describing the fit of the obtained model to the particular data being analyzed, these estimators are "too high" in the sense that they tend to overestimate how well this model would fit the population. The bias arises because measurement and sampling errors, as well as true variance, are considered in determining the coefficients which provide maximum explanatory power in the data being analyzed.

If one assumes that a fixed-effects model holds for $\eta_{i\ell}^2$ and that the multiple regression assumptions are met in MNA, unbiased estimators can be calculated in place of the $\eta_{i\ell}^2$'s and the R_ℓ^2's (McNemar, 1962, pp. 184-185). As noted in Chapter 3, MNA presently computes and prints the unbiased--i.e. "adjusted"--versions of the R_ℓ^2's. What is needed are the unbiased counterparts for the generalized eta-square, η_i^2, and the generalized R-square, $\mathcal{R}^2$.

A simple substitution of the unbiased version of the $\eta_{i\ell}^2$'s, S_ℓ^2's, and R_ℓ^2's is not satisfactory. This is because they appear in products in generalized expressions (see equations 3-24, 3-29) and the product of expectations is in general not equal to the expectation of a product. Ideal statistics are minimum variance unbiased estimators to replace η_i^2 and $\mathcal{R}^2$.

Analysis and Monte Carlo simulation are planned to investigate this area.

6.3 Testing for Statistical Significance

As mentioned, the authors' research philosophy leans towards strength of relationship statistics and away from tests of statistical significance.* In addition, the survey data for which MNA was written are frequently weighted and often arrived at through stratified sampling, which make the interpretation of significance tests difficult if not impossible. For these reasons, the MNA output does not contain any significance tests.

The most common significance tests, namely

$$Ho: \quad \eta^2_{i\ell} = 0$$

$$Ho: \quad R^2_{\ell} = 0$$

are well documented and could be made using F statistics (Hays, 1963, p. 368-370, 573). At present, however, the MNA output does not contain the within dependent variable code frequency distributions across independent variable codes necessary to compute the F-statistics.

It would be of interest to know the procedure for significance testing the generalized version of the above statistics, that is the tests

$$Ho: \quad \eta^2_{i} = 0 \text{ and}$$

$$Ho: \quad \mathcal{R}^2 = 0.$$

Work is proposed to determine the distributions of η^2_{i} and $\mathcal{R}^2$ using both analytical and Monte Carlo methods.

In addition, there is interest in a significance test for the marginal contribution of the ith independent variable to $\mathcal{R}^2$, that is

$$Ho: \Delta \mathcal{R}^2_{i} = 0 \quad \text{where } \Delta \mathcal{R}^2_{i} \text{ is the difference in } \mathcal{R}^2 \text{ be-}$$

fore and after adding predictor i to the model. Another test of interest

---

*For a discussion of the relative merits of significance testing and predictive strength see Hays (1963, p. 299-300) and Morrison and Henkel, (1970).

is whether the reduction in R^2 due to the additive model is significantly different from 0, that is if we define $\tilde{R}^2$ as the generalized R^2 value using all predictors in a pattern variable (see Chapter 5) then $\tilde{R}^2 \geq R^2$ and the difference or "interaction term"

$$\Delta \ \tilde{R}^2 = \tilde{R}^2 - R^2$$

can be considered as the amount of potential explanatory power lost due the additivity assumption.

Of secondary interest is a parallel investigation of the θ_i and Θ statistics. The θ_i distribution should be obtained with relative ease as it is a linear transformation of the Goodman and Kruskal λ statistic (see Chapter 3).

6.4 Residuals

Because of the proven usefulness of analyzing residual scores derived from intervally scaled dependent variables, it was decided to include a residuals option in MNA on an experimental basis in an attempt to generalize the residuals methodology to nominal dependent variables. Comments on use and description of output is found in Chapter 5 and Appendix A.

One feature which is important to keep in mind with respect to residual scores from MNA is that there is a set of residual-score variables, not just a single variable. There is one residual-score variable for each category of the original nominally scaled dependent variable. As with residual scores from other multivariate techniques, the residual scores from MNA express the deviation of each object from the prediction for that object. The prediction we have chosen to use is the MNA-derived probability of the object being classified in a particular category of the dependent variable. For example, a person with a particular combination of characteristics might have a likelihood of .31 of being a pacifist, .37 of being a vigilante, .13 of being an anarchist, and so forth.* (Of course, the probabilities will add to 1.00 if all the categories are included.) The residual scores are the differences between the MNA-derived probabilities and 0 or 1, depending on whether or not the object actually did fall in the designated category of the dependent variable.

6.4a Uses for MNA residuals. One of the relatively unexplored aspects of MNA is the use of the set of residual-score variables which can be derived from it. Following are two least squares regression residual techniques the authors feel need investigation concerning their applicability to MNA:

A. Stagewise least-squares (Draper and Smith, 1966, pp. 173-177):

Here the effect of one set of predictor variables is removed from the dependent variable and the resulting residual scores, used as new dependent variables, are analyzed using a different predictor set. This process can be carried out to any desired stage and allows the user control over the order in which predictor variables explain variance.

Each of the residual-score variables is a numerical variable, and as such might appropriately be used in multiple regression, multiple classi-fication analysis, analysis of variance, and other multivariate techniques making similar assumptions about dependent variables. For some purposes one might wish to subject just one of the residual-score variables output by MNA to subsequent analysis--e.g., to try to understand why people did or did not have the orientations of a pacifist. However a complete analy-

---

*These figures are actual data taken from the sample MNA run included in Appendix C for persons who are white, have a college degree, and who grew up in New England--see Table 2.2. These people also have probabilities of .13 of falling in the "inter-mediate" class, and of .06 of falling in the "warrior" class.

sis would focus on the whole <u>set</u> of residual variables.

Unfortunately techniques for combining results from sets of numerical dependent variables are complex and sometimes do not lend themselves to simple interpretation. Use of canonical correlation and multivariate analysis of variance may be possible. Application of the generalized R^2 described in this monograph may also be possible. We suspect, however, that analysts would be well advised to seek to understand the results of analyzing each residual-score variable <u>separately</u> before attempting to combine these variables into one global analysis.

B. Inspection of outlying residuals:

The objective here is to identify cases having unusually large residual scores. While such scores may be the result of strong real effects, they may also be the result of some error in data preparation. Identifying such cases should facilitate checking for errors. It would not be difficult to develop a scanning algorithm which would indicate sets of residual scores beyond a critical standard error value.

<u>6.4b Calculation of residual scores</u>. A second general problem having to do with residuals is determination of precisely what residual score, or set of scores, is most useful. As noted above we have chosen to output sets of scores which are deviations of predicted likelihoods from actuality. Another alternative would have been to put out a set of residuals (one for each category of the original nominally scaled dependent variable) which would express in simple 0 - 1 fashion whether or not the object did fall in the category predicted on the basis of its <u>forecast</u>. An object which fell in the category predicted would receive scores of "0" on <u>each</u> of the residual-score variables--i.e. no deviation from expectation. An object

which fell in a category other than that predicted would receive two "1's" among its set of residual scores--one in the variable for the category it was predicted to fall into, but did not, and one in the variable for the category it did fall into, but was not predicted to. (If there were more than two categories in the original dependent variable, the other residual score variables would be scored zero.) We have not experimented with this second method of scoring residuals. It is an interesting alternative, however, because it incorporates the type of decision rule which is at the basis of much nominal scale analysis, including the multivariate Theta statistic which is output by MNA.

Having described this second method of calculating residuals, still a third method comes readily to mind. This would be a single variable residual score indicating simply whether the object did or did not fall in the category predicted for it. This would represent a kind of collapse of the variables described in the second method. In the process of collapsing across variables, information about the location of "errors" would be lost. We feel this is less promising than either of the previously described methods, though useful applications for such a summary measure might be found.*

6.5 Importance of Predictors

A general issue in any multivariate analysis is the importance of the several predictors in explaining variation in the dependent variable. This issue has received much attention over the years, though most of the discussions focus on analyses involving an intervally scaled dependent variable, and in particular, on the technique of multiple regression. A

---

*Note that with the last two definitions of residual scores it would be possible to do a stagewise analysis using MNA as the analysis technique for all stages.

recent article by Darlington (1968), for example, reviews five different measures of the importance of a predictor, and endorses three. The purposes of this section are to briefly summarize the different criteria used when assessing the importance of predictors and mention some of the measures which are in current use with respect to multiple regression (Section 6.5a), and then to extend the discussion to multivariate analyses of nominally scaled dependent variables (in Sections 6.5b and 6.5c).

6.5a Three criteria for assessing importance and measures conventionally used. There are three distinct criteria which analysts use when assessing the importance of predictors.

One criterion is the simple relationship between the dependent variable and the predictor. (We designate this Criterion I.) Adopting this criterion, one asks how well a particular predictor, considered alone, can explain variation in the dependent variable. Any bivariate statistic appropriate to the data provides a measure for assessing the importance of predictors using this criterion. In data to which multiple regression is conventionally applied, this bivariate statistic would usually be the Pearson product-moment correlation coefficient. The predictor having the highest correlation with the dependent variable, is, using this criterion, the most important.

Criterion I, while useful for some purposes, does not consider the fact that predictors may be correlated among each other and hence may "overlap" in their explanatory power. (This phenomenon is sometimes referred to as "multi-collinearity.") For example, even though Predictors A and B might each be more highly related to a dependent variable than Predictor C, Predictors A and C might together provide better predictions of the dependent variable than would Predictors A and B jointly. (This is what one would expect if Predictors A and B were themselves highly correlated and neither was much correlated with C, and if C bore some relationship to the dependent variable.) Thus, a second criterion for importance is the usefulness of the predictor in increasing one's ability to explain a dependent variable. The basic notion here is that of "marginal" or "unique" explanatory power--how much of the variation in a dependent variable can a particular predictor explain over and above what can be explained by other predictors? (We call this Criterion II.)

In conventional multiple regression this marginal predictive power is assessed by several different measures. One approach is to determine the amount by which the squared multiple correlation increases when the predictor in question is added to the analysis. (E.g., one computes the percentage of variance explained in a dependent variable by Predictors A and B, and also by A, B, and C; and the difference in the squared multiple correlation coefficient is the marginal predictive power of Predictor C.) This difference is what Darlington (1968, p. 162) defines as a predictor's "usefulness". It is equivalent to the squared part correlation (sometimes referred to as the squared semi-partial correlation) between the dependent variable and the predictor in question.* This measure assesses the importance of a predictor in terms of the variance in the dependent variable marginally explainable by the predictor, relative to the total variance in the dependent variable.

Another measure widely used to assess the marginal usefulness of a predictor is the partial correlation. The partial correlation measures the variance in the dependent variable marginally explainable by the pre-

---

*Actually there are two possible part correlations. The relevant one for this discussion has the effects of other predictors removed from the predictor whose importance is being assessed but not from the dependent variable.

dictor relative to the as-yet-unexplained variance in the dependent variable. Within any one set of predictors, the partial correlations will rank order the importance of predictors in a manner identical with the rank order obtained from the part correlations. However, the numerical values of part and partial correlations are not the same, and they are not monotonically related across predictors which come from different sets.

The third criterion used to assess the importance of a predictor asks how the predictor relates to the dependent variable if all other variables are "held constant"--i.e., if, in each subgroup of the predictor in question, all other predictors were distributed as they are in the population at large. (We call this Criterion III.) In conventional multiple regression, this criterion is assessed by the standardized regression coefficient, often denoted as Beta.

While it is often true that standardized regression coefficients and partial (or part) correlations rank order the importance of predictors within a set similarly, this is not necessarily the case (Ezekiel and Fox, 1966, p. 197). The relationships between Beta and the partial correlation depend on the correlations among the predictors, and between predictors and the dependent variable:

$$r^2_{XA \cdot B \ldots n} = \beta^2_A \left(\frac{1 - R^2_{A \cdot B \ldots n}}{1 - R^2_{X \cdot B \ldots n}} \right)$$

$$X = \text{dependent variable}$$
$$A, B \ldots n = \text{independent variables.}$$

Beta will exceed the partial correlation whenever the predictor in question is itself more predictable from the other predictors than is the dependent variable.

6.5b <u>Assessing importance of predictors in MNA</u>. While measures for assessing predictor importance under the above three criteria are reasonably
well worked out with respect to multiple regression, things are more difficult when we move to multivariate analysis of categorical dependent variables. In principle the three criteria sketched above are as applicable
to categorical dependent variables as they are to intervally scaled ones.
However, we lack simple ways of assessing importance under Criteria II and
III.

For Criterion I, we believe the bivariate Theta and generalized eta
statistics produced by MNA provide useful ways of assessing predictor
importance.

With respect to Criterion II, we know of no statistical formula by
which one could compute, on the basis of a single MNA run, the analogue
of either the part or partial correlations so useful in multiple regression
for assessing marginal importance of a predictor. While someone may
eventually develop an appropriate formula, for the moment the only way to
assess the exact marginal importance of a predictor is to run MNA twice--
once including the predictor and once omitting it--and observe the change
in the multivariate Theta or generalized R^2 statistics. This procedure,
while cumbersome, does offer a precise way of assessing a predictor's
marginal importance in a given data set.*

Assessing importance under Criterion III--the importance of a predictor after "holding constant" all other predictors--also presents prob-

---

*It would also be desirable to be able to estimate the marginal
importance of a predictor in the population at large. Logically
this would involve examining differences in squared multiple
correlations which had been adjusted for degrees of freedom.
However, as discussed in Section 6.2, we do not yet know an
appropriate way to make this adjustment in the MNA context.

lems.   Instead of having a single standardized regression coefficient for each predictor, as one does in multiple regression, in MNA the effect of each predictor after holding constant all other predictors is assessed by a two dimensional array of "effect" coefficients--the a's described in Chapter 3.   These can be interpreted as showing the effect of membership in each category of each predictor on the probability of a case falling in a particular category of the dependent variable, after "holding constant" the effects of all other predictors.   While these effect coefficients do not, by themselves, provide an easily interpretable measure of importance, it is nevertheless true that, in general, predictors with the biggest effects are the ones we would wish to consider most important.*   The Beta statistic produced by MNA is an attempt to capitalize on this observation.

6.5c   The Beta statistic.   The Beta statistics included in the MNA output provide a partial solution to the problem of assessing the importance of predictors under Criterion III.   For each predictor, there is one Beta computed for each category of the dependent variable.   These Betas are comparable across the various independent variables, and across the various categories of the dependent variable.

    The Betas are useful for exploring two kinds of questions: 1) the relative predictive power of different independent variables with respect to a given category of the dependent variable, and (2) the relative predictive power of a given independent variable with respect to the differ-

_____

*Of course, this would be true only after allowing for differences between predictors in the distribution of cases across their categories.   For example, a predictor having one category which was sharply different from the others with respect to the dependent variable, but which had only a small proportion of the cases in that distinct category, might not be as important as another predictor with smaller effects involving a larger number of cases.

ent categories of the dependent variable, in both cases within the context of "holding other variables constant."

These category-specific Betas are based on the effect coefficients discussed above and provide a convenient and standardized way of summarizing their magnitudes and at the same time taking into account the number of cases to which each of the coefficients applies. (Chapter 3 provides the formula by which Beta is computed.) They are the same as those produced by the Multiple Classification Analysis Program described by Andrews, Morgan, and Sonquist (1967, pp. 117-122). The reader is referred to that document for a fuller description of the nature of Beta and why it seems closely analagous to the standardized partial regression coefficient from conventional multiple regression.

While the Betas provide a partial solution to the problem of assessing predictor importance under Criterion III, they are specific to a particular category of the dependent variable. At times it would be convenient to have a summary measure which would indicate the importance of predictors, holding others constant, in accounting for membership in all categories of the dependent variable. Such a statistic would be a "generalized Beta" and would be analogous to the generalized etas and generalized multiple correlation presently produced by MNA. Although Chapter 3 presents a formula for an experimental version of a generalized Beta, this statistic is not presently produced by the program. The development and interpretation of such a statistic await further exploration.

Bibliography

Andrews, F.M., Morgan, J.N., and Sonquist, J.A. The Multiple Classification Analysis Program: A Report of a Computer Program. Ann Arbor, Michigan, Institute for Social Research, 1967.

Blumenthal, M.D., Kahn, R.L., Andrews, F.M., and Head, K.B. Justifying Violence: Attitudes of American Men. Ann Arbor, Michigan, Institute for Social Research, 1972.

Bohrnstedt, G.W. and Carter, T.M. Robustness in regression analysis. In: Costner, H.L. (Ed.), Sociological Methodology 1971. San Francisco, Jossey-Bass, 1971.

Coleman, J.S. Introduction to Mathematical Sociology. New York, Macmillan, 1964.

Coleman, J.S. Multivariate analysis for attribute data. In: Borgatta, E.F. and Bohrnstedt, G.W. (Eds.), Sociological Methodology 1970. San Francisco, Jossey-Bass, 1970.

Cooley, W. W. and Lohnes, P.R. Multivariate Data Analysis. New York, Wiley, 1971.

Darlington, R.B. Multiple regression in psychological research and practice. Psychological Bulletin, 1968, 69, 161-182.

Draper, N. and Smith, H. Applied Regression Analysis. New York, Wiley, 1966.

Ezekiel, M. and Fox, K.A. Methods of Correlation and Regression Analysis. (3rd ed.) New York, Wiley, 1966.

Fisher, R.A. The use of multiple measurements in taxonomic problems. Annals of Eugenics, 1936, 7, 179-188.

Forsythe, G. and Moler, C.B. Computer Solution of Linear Algebraic Systems. Englewood Cliffs, New Jersey, Prentice Hall, 1967.

Goodman, L.A. On partitioning chi square and detecting partial association in three-way contingency tables. Journal of the Royal Statistical Society, Series B, 1969, 31, 486-498.

Goodman, L.A. The multivariate analysis of qualitative data: interactions among multiple classifications. Journal of the American Statistical Association, 1970, 65, 226-256.

Goodman, L.A. Partitioning of chi square, analysis of marginal contingency tables, and estimation of expected frequencies in multidimensional contingency tables. Journal of the American Statistical Association, 1971, 66, 339-344.

Goodman, L.A. A modified multiple regression approach to the analysis of dichotomous variables. The American Sociological Review, 1972a, 37, 28-46.

Goodman, L.A. A general model for the analysis of surveys. American Journal of Sociology, 1972b, 77, 1035-1086.

Goodman, L.A. and Kruskal, W.H. Measures of association for cross classifications. Journal of the American Statistical Association, 1954, 49, 732-64.

Grizzle, J.E., Starmer, C.F., and Koch, G.G. Analysis of categorical data by linear models. Biometrics, 1969, 25, 489-504.

Hays, L. Statistics for Psychologists, New York, Holt, Rinehart and Winston, 1963.

Irwin, J.O. A note on the subdivision of X^2 into components. Biometrika, 1949, 36, 130-134.

Labovitz, S. The assignment of numbers to rank order categories. American Sociological Review, 1970, 35, 515-24.

Lancaster, H.O. The derivation and partition of X^2 in certain discrete distributions. Biometrika, 1949, 36, 117-129.

Lancaster, H.O. Complex contingency tables treated by the partition of X^2. Journal of the Royal Statistical Society, Series B, 1951, 13, 242-249.

McNemar, Q. Psychological Statistics (3rd Edition), New York, Wiley, 1962.

Messenger, R.C. and Mandell, L.M. A modal search technique for predictive nominal scale multivariate analysis. Journal of the American Statistical Association, 1972, 67, 768-772.

Miller, R.G. Statistical prediction by discriminant analysis. Meteorological Monographs, 1962, 4(25).

Miller, R.G. Regression estimation of event probabilities. Technical Report 7411-121, Contract Cwb 10704. Hartford, Conn., The Travelers Research Center, Inc, 1964.

Morgan, J.N. and Messenger, R.C. THAID. Ann Arbor, Michigan, Institute for Social Research, 1973.

Morrison, D.E., and Henkel, R.E. (Eds.) The Significance Test Controversy. Chicago, Aldine, 1970.

Nunnally, J.C. Psychometric Theory, New York, McGraw-Hill, 1967.

O'Malley, P. An empirical comparison of MNA and MDF. Unpublished paper, Department of Psychology, University of Michigan, 1972.

Overall, J.E., and Klett, C.J. _Applied Multivariate Analysis_. New York, McGraw-Hill, 1972.

Sonquist, J.A. _Multivariate Model Building_. Ann Arbor, Michigan, Institute for Social Research, 1970.

Sonquist, J.A., Baker, E.L., and Morgan, J.N. _Searching for Structure_. Ann Arbor, Michigan, Institute for Social Research, 1971.

Stevens, S.S. On the theory of scales of measurement. _Science_, 1946, _103_, 677-680.

Suits, D.B. Use of dummy variables in regression equations. _Journal of the American Statistical Association_, 1957, _52_, 548-551.

Sutcliffe, J.P. A general method of analysis of frequency data for multiple classification designs. _Psychological Bulletin_, 1957, _54_, 134-137.

Theil, H. On the estimation of relationships involving qualitative variables. _American Journal of Sociology_, 1970, _76_, 103-154.

University of Michigan, Institute for Social Research. _OSIRIS III Volume 1: System and Program Description_. Ann Arbor, Michigan, The University of Michigan, 1973a.

University of Michigan, Institute for Social Research. _OSIRIS III Subroutine Manual_. Ann Arbor, Michigan, Institute for Social Research, 1973b.

Walker, S.H. and Duncan, D.B. Estimation of the probability of an event as a function of several independent variables. _Biometrika_, 1967, _54_, 167-179.

Appendix A

SET-UP INSTRUCTIONS

MNA is part of the Institute for Social Research's OSIRIS package of computer programs. The control cards and method of access therefore conform to OSIRIS format. Users should consult OSIRIS III Volume 1: System and Program Description (University of Michigan, 1973a) and The OSIRIS III Subroutine Manual (University of Michigan, 1973b) for details.

Input

    1. Dictionary, on cards, tape, or disk.

    2. Data, on cards, tape, or disk.

    3. Control cards.

Output

    1. Number of cases passing global filtering.

    2. Parameter interpretation.

    3. Unweighted frequency.

    4. Weighted frequency.

    5. Number of local-filtered cases.

    6. Number of cases outside valid ranges.

    7. Non-empty predictor codes.

    8. Minimum number of significant digits in solution vectors.

    9. Univariate distribution of dependent variable--unweighted, weighted, and percentized weights.

   10. Predictor summary:

       a. Univariate distribution of predictor

          i) Frequency.

          ii) Weighted-frequency.

iii) Percentized weighted-frequency.

b. Bivariate Statistics:

i) Weighted frequency percentized marginals.

ii) Adjusted means.

iii) Coefficients.

11. Summary Statistics:

a. Individual predictors:

i) Theta

ii) Eta-squared (each dependent-variable code).

iii) Beta-squared (each dependent-variable code).

iv) Generalized eta-squared.

b. Joint Prediction:

i) R-squared (each dependent-variable code).

ii) Adjusted R-squared (each dependent variable code).

iii) Generalized R-Squared.

iv) Joint Theta (proportion correctly classed).

12. Classification Matrix:

(Rows indicate actual codes, columns indicate predicted codes).

13. Residuals (optional)--up to ten eight-column residual fields (one) for each dependent-variable code), preceeded by the ID. Variable.

Items 2-13 are repeated for each analysis packet. Time is printed after the input data is read, after generation of the sums-of-squares/ cross-products matrix, after solution of the linear system, and after calculation and printing of the summary statistics (in that order).

Restrictions

1. Maximum number of non-empty dependent variable codes = 10.

2. Range of dependent variable: 0-9.

3. Maximum number of non-empty codes for any independent variable: 20.

4. Range of independent variables: 0-19.

5. Maximum sum of non-empty independent variable codes: 100.

6. Number of analysis packets is unlimited.

7. Maximum number of variables in global variable list plus unlisted recode and/or filter variables: 100.

8. Any variable used as a weight variable or local-filter variable must be in the global variable list.

Missing-data Treatment

Missing-data elimination is specified by the local keyword parameter MDOPT.

Setup Summary

JOB card

// EXEC OSIRIS

//DICTx   DD parameters defining input dictionary (omit if on cards)

//DATAx   DD parameters defining input data (omit if on cards)

//SETUP   DD *

$RUN MNA

$RECODE card ⎫
             ⎬ optional   (See University of Michigan, 1073a)
Recode statements ⎭

$SETUP card

A. Global-filter card (optional).

B. Label card.

C. Main parameter card.

D. Global variable list.

E. Local-filter card (optional).

F. Local label card.

G. Local parameter card.

H. Name card for dependent variable codes (optional).

I. Local variable-list card.

$DICT card

  Dictionary-descriptor card   If dictionary on cards.

  T-cards

                         (See University of Michigan, 1973a)

$DATA card

  Data cards   If data on cards.

/* card

(Note: Control cards E-I form an "Analysis Packet" and must be repeated
for each packet.)

Description of Control Cards

A. <u>Global-filter</u> <u>card</u> (optional). Filter cards select a subset of
cases by specifying certain values of certain variables. A
"global" filter applies to the entire program run. A "local"
filter applies only to one computation (e.g., one table in an
analysis program) performed during the program run.

Examples: (a) INCLUDE V2=1-5 AND V7=23,27,35, AND V8=1,3,4,6*

             (b) EXCLUDE V10=2-3,6,8-9 AND V30=001-004 OR V91=025*

1. Punch INCLUDE or EXCLUDE, beginning anywhere on the first
card. Continue punching on the first card; if necessary extend by ending the first card with a comma or conjunction (AND
or OR) and continuing on the next card.

2. Maximum number of expressions per run: 15 (an expression
includes V, the variable number, an equals sign, and a list of

values).

3. An asterisk must terminate the list of expressions.

4. Variables may appear in any order and in more than one expression.

5. Expressions are connected by AND or OR:

    a. AND indicates that the values in all connected expressions must be found in order to select the case.

    b. OR indicates the case will be selected if any or all of the specified values are found.

    c. AND expressions are executed before OR expressions.

    d. Example: expression 1 AND expression 2 OR expression 3. A case would be selected if any expression 3 values were found or if all expression 1 and 2 values were found, but not if only expression 1 values were found or if only expression 2 values were found.

6. Values specified must have the field width of the pertinent variable; lead zeroes must be punched.

7. Values may be specified singly, separated by commas, or in a range (e.g., 001-004 using a dash).

8. Values may be positive or negative, but a value range may not vary from a negative value to a positive value (separate into two ranges).

9. Negative ranges should be expressed as in example (a) for global filter, as in (a) or (b) for local filter: (a) V1=-10--1,0-9*

    (b) V1=-1--10,0-9*

10. Error messages are self-explanatory; e.g., EQUAL SIGN NOT PRECEDED BY A V or INCONSISTENT FIELD WIDTH.

B. <u>Label</u> <u>card</u>--up to 80 characters, punched free-form, to title the printout.

C. <u>Main</u> <u>Parameter</u> <u>card</u>. Keywords (defaults underlined) are separated by commas or blanks; the list is terminated by an asterisk. If all defaults are chosen, punch a single asterisk on parameter card.

INFILE=<u>IN</u>/xxxx

Allows the user to specify a 1-4 character ddname suffix.

<u>DICT</u>/NODICT

Print the input dictionary?

BADDATA=<u>TERMINATE</u>/STOP/MD1/MD2/SKIP

When non-numeric characters (including embedded blanks and all-blank fields) are found in numeric variables:
SKIP: Skip the case.
MD1: Convert the value to MD1 code
MD2: Convert the value to MD2 code
<u>TERM</u>: Terminate the run.
STOP: Terminate the run.
For SKIP, MD1, MD2: a message reports the number of cases so treated.

(Note: MD1 and MD2 refer to "Missing Data Types" -- See University of Michigan, 1973a)

D. <u>Global</u> <u>Variable-list</u> <u>card</u>. Variable-list cards are used to specify certain variables to be used in a program run.

Examples:

(a) V1-V6,V9,V16,V20-V102,V18,V11,V209*

(b) V2,V5,V7,V10,V12,V15,V21,V26,V29,V31-V45, V52,V67-V92,V115,V136* (Continuation Card)

1. Columns 1-80 may be used.

2. Punching is free format (i.e., blanks anywhere).

3. An asterisk must follow the last item on the last card.

4. Variables may be specified in any order.

5. Single variables or variable ranges (e.g., V31-V45) are separated by commas.

6. V must precede every variable number (e.g., V31-45 is not valid).

7. Continuation: stop with a comma, and continue punching anywhere on the continuation card; use as many continuation cards as needed.

8. Errors are reported by the message, ERROR IN VARIABLE LIST N, where N is one of the following:

    1--First character not V.

    2--Comma not followed by V.

    3--Decreasing sequence in a variable range (e.g., V10-V5).

    4--V misplaced or missing comma.

    5--Last character on card not comma or asterisk.

    6--Variable number greater than 9999.

    7--Asterisk misplaced.

    8--Dash misplaced.

E. Local filter card: See A.

F. Local label card--up to 80 characters, punched free-form, to title the printout.

G. Local parameter card: Keywords are chosen from those described below (defaults are underlined), are separated by blanks or commas, and terminated with an asterisk.

    DEPVAR=n   Dependent-variable number. (No default.)

    WEIGHT=n   Weight-variable number. (Default: no weights.)

MDOPT=<u>NONE</u>/MD1/MD2/BOTH	What dependent variable miss-ing-data cases should be <u>eliminated?</u>
RESIDUALS/<u>NORESIDUALS</u>	Residuals to be computed, printed and written on file DATAOUT.
IDVAR=n	ID variable number for resid-uals output. (No Default). (Must appear in both global and local variable lists).
<u>SUPPRESS</u>/NOSUPPRESS	Suppress printout of residuals?
<u>NONAME</u>/NAME	Names supplied for the depen-dent-variable codes?
RSCALE=n/3	Scale factor for <u>residuals</u> output (only when residuals requested). Residuals mul-tiplied by 10 raised to the Rscale power.
OUTFILE=<u>OUT</u>/xxxx	Output ddname suffix for <u>residuals</u> output.

H. <u>Name card</u>: If the keyword NAME is specified on the local para-meter card, this card specifies up to 10 continuous eight-column fields used to title the non-empty dependent-variable codes in order.

I. <u>Local variable-list card</u>: See D.

Error Messages

(Errors marked with an asterisk (*) terminate the run).

1. *GLOBAL PARAMETER CARD MISSING

2. *ERROR IN GLOBAL PARAMETER CARD

3. *LOCAL PARAMETER CARD MISSING

4. *ERROR IN LOCAL PARAMETER CARD

5. *NO DEPENDENT VARIABLE SPECIFIED

6. *VARIABLE SPECIFIED BUT NOT INCLUDED IN THE LOCAL VARIABLE LIST

7. *DUMMY PREDICTORS — EXCEEDING LIMIT OF 100

8. *MATRIX WITH ZERO ROW IN DECOMPOSE

9. *SINGULAR MATRIX IN DECOMPOSE. ZERO DIVIDE IN SOLVE

10. **** WARNING **** MATRIX DOES NOT CONVERGE

Notes: Message 8 occurs only if an initial predictor is a constant for the data being analyzed.

Message 9 occurs when a linear dependency exists in the sums-of-squares — cross products matrix (See Chapter 5 — Restrictions).

Message 10 indicates an extremely "ill-conditioned" matrix preventing solution of the linear system to the level set internally (See Chapter 5 — Program Design).

94

MNA MACRO FLOW CHART

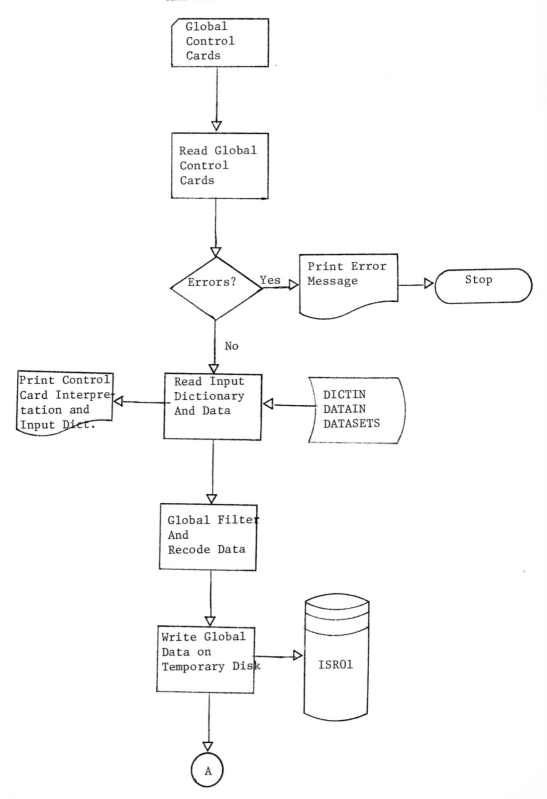

95

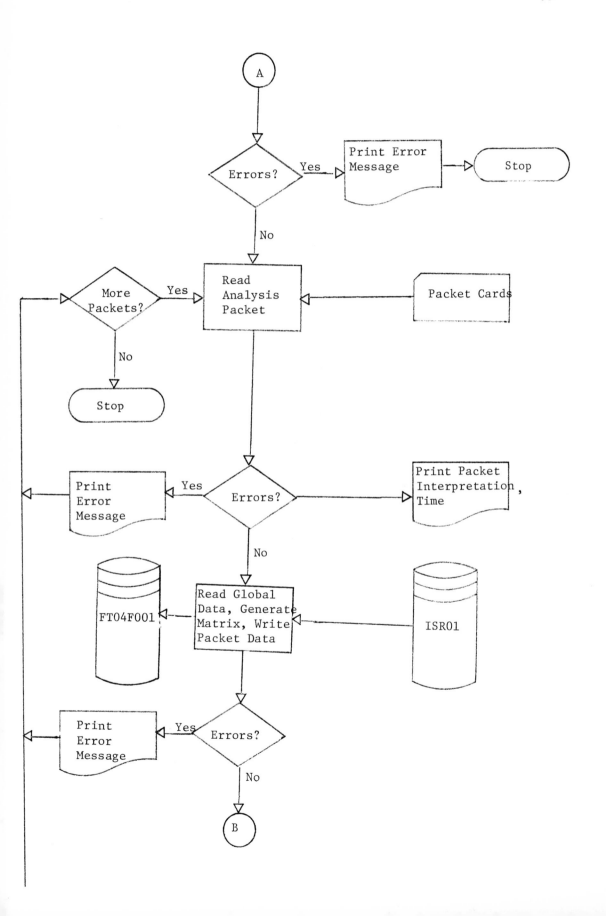

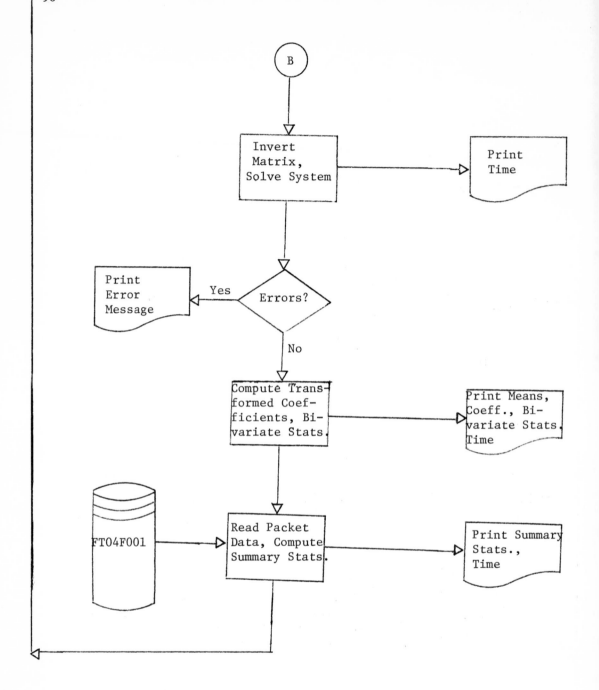

APPENDIX C

SAMPLE OUTPUT

JOB 166

```
//MO12767 JOB (,
// 468473,MNA,2,...,5),MESSENGER,MSGLEVEL=(1,0)                        00000020
// EXEC ISRSYS                                                         00000040
XXCSIRIS PROC LIB=OSIRPGM,LIB1=SRCLIB,LIB2=CPSLIB,LIB3=ISRLIB,         00000060
XX          DI=**.FT48F001,VOL=REF=*.FT48F001',                        00000080
XX          DA=**.FT47F001,VCL=REF=*.FT47F001',                        00000100
XX          P=ISRSYS,SP1=1000,SP2=600
XXGO   EXEC  PGM=&P                                                     00000120
IEF6531 SUBSTITUTION JCL - PGM=ISRSYS
XXSTEPLIB CD DSN=&LIB,DISP=SHR                                          00000140
IEF6531 SUBSTITUTION JCL - DSN=OSIRPGM,DISP=SHR
XX        DD DSN=&LIB1,DISP=SHR                                         00000160
IEF6531 SUBSTITUTION JCL - DSN=SRCLIB,DISP=SHR
XX        DD DSN=&LIB2,DISP=SHR                                         00000180
IEF6531 SUBSTITUTION JCL - DSN=CPSLIB,DISP=SHR
XX        DD DSN=&LIB3,DISP=SHR
IEF6531 SUBSTITUTION JCL - DSN=ISRLIB,DISP=SHR
XXSYSPUNCH DD SYSOUT=B                                                  00000200
XXSYSPRINT DD SYSOUT=A                                                  00000220
XXSYSOUT   DD SYSOUT=B                                                  00000240
XXFT02F001 DD SYSOUT=B                                                  00000260
XXFT03F001 DD UNIT=SYSDA,SPACE=(TRK,(100,50)),                         00000280
XX  DCB=(RECFM=VBS,LRECL=204,BLKSIZE=204,BUFNO=1)                       00000300
XXFT04F001 DD UNIT=SYSDA,SPACE=(TRK,(200,50))                          00000320
XX  CCB=(RECFM=VBS,LRECL=200,BLKSIZE=2C4,BUFNO=1)                       00000340
XXFT05F001 DD UNIT=SYSDA,SPACE=(TRK,(50,50)),                          00000360
XX  DCB=(RECFM=VBS,LRECL=200,BLKSIZE=204,BUFNO=1)                       00000380
XXFT06F001 CD SYSOUT=A                                                  00000400
XXFT07F001 DD UNIT=SYSCA,SPACE=(TRK,(50,50)),                          00000420
XX  DCB=(RECFM=VBS,LRECL=200,BLKSIZE=204,BUFNO=1)                       00000440
XXFT08F001 DD UNIT=SYSCA,SPACE=(TRK,(50,101)),                         00000460
XX  CCB=(RECFM=FB,LRECL=80,BLKSIZE=80)                                  00000480
XXFT46F001 DD UNIT=SYSCA,SPACE=(TRK,(5,51)),                           00000500
XX  CCB=(RECFM=FB,LRECL=80,BLKSIZE=80C)                                 00000520
XXFT47F001 DD UNIT=SYSCA,SPACE=(TRK,(100,20)),                         00000540
XX  CCB=(RECFM=FB,LRECL=80,BLKSIZE=3520)                                00000560
XXFT48F001 DD UNIT=SYSCA,SPACE=(TRK,(100,20)),                         00000580
XX  CCC=(RECFM=FB,LRECL=8C,BLKSIZE=3520)                                00000600
XXFT49F001 DD UNIT=SYSCA,SPACE=(TRK,(200,50)),                         00000620
XX  CCB=(RECFM=FB,LRECL=8C,RLKSIZE=3520)                                00000640
XXFT50F001 DD DSN=ISRNEWS,DISP=SHR,LABEL=(,,,IN)                       00000660
XX   VOL=REF=*.FT47F001,DSN=*.FT47F001,DISP=(OLD,DELETE)                00000680
XXISRO1 DD UNIT=SYSDA,SPACE=(TRK,(ESP1),,CCNTIG)                       00000700
IEF6531 SUBSTITUTION JCL - UNIT=SYSDA,SPACE=(TRK,(1000),,CONTIG)
IEF6531 SUBSTITUTION JCL - UNIT=SYSDA,SPACE=(TRK,(600),,CONTIG)         00000720
XXISRO2 DD UNIT=SYSDA,SPACE=(TRK,(ESP2),,CCNTIG)
XXISRO3 DD UNIT=SYSDA,SPACE=(TRK,(50),,CCNTIG)                          00000740
XXISRO9 DD UNIT=SYSDA,SPACE=(TRK,(200),,CONTIG)                         00000760
XXISR1Q DD UNIT=SYSDA,SPACE=(TRK,(200),,CONTIG)                         00000780
XXUCLOAC DD DSN=*.FT07F001,VOL=REF=*.FT07F001,DISP=(OLD,DELETE)         00000800
XXSORTWK01 DD DSN=*.ISRO1,DISP=(OLD,DELETE),VOL=REF=*.ISRO1             00000820
XXSORTWK02 CD DSN=*.ISRO2,DISP=(OLD,DELETE),VOL=REF=*.ISRO2             00000840
XXSOPTWK03 CD DSN=*.ISRO9,DISP=(OLD,DELETE),VOL=REF=*.ISRO9             00000860
XXSORTWK04 CD DSN=*.ISR1O,DISP=(OLD,DELETE),VOL=REF=*.ISR1O             00000880
XXSORTWK05 CD DSN=*.ISRO3,DISP=(OLD,DELETE),VOL=REF=*.ISRO3             00000900
XXSOPTWK06 CD DSN=*.FT04F001,DISP=(OLD,DELETE),VOL=REF=*.FT04F001       00000920
XXSORTLIB DD DSN=SYS1.SORTLIB,DISP=SHR                                  00000940
XXSORTIN  DD VOL=REF=*.FT47F001,DSN=*.FT47F001,DISP=(OLD,PASS)          00000960
XXSOPTCOUT DC VOL=REF=*.FT47F001,DSN=*.FT47F001,DISP=(OLD,PASS),
XX  CCB=(RECFM=FB,LRECL=80,BLKSIZE=80C)                                 00001000
//DICTTIN DD DSN=DICT,VIOL8,UNIT=T,VOL=SER=3180,LABEL=1,DISP=OLD
```

THIS IS A GENERAL PURPOSE PROCEDURE FOR OSIRIS PROGRAMS.

DISK FILE FOR LOCAL DATA. REQUIRES 4 BYTES/ VARIABLE FOR EACH CASE.

DISK FILE FOR GLOBAL DATA. REQUIRES 4 BYTES/ VARIABLE FOR EACH CASE.

INPUT DICTIONARY (INDICATES VARIABLE CHARACTERISTICS)

```
X/CICTIN  DD  DSN=&DI,DISP=(OLD,PASS),                                         00001020
IEF653I SUBSTITUTION JCL - DSN=*.FT48FOO1,VOL=REF=*.FT48FOO1,DISP=(OLD,PASS),
XX   DCB=BUFNO=1                                                               00001040
X/DATAIN  DD  DSN=DATA.VIOL8,UNIT=T,VOL=SER=3180,LABEL=2,DISP=OLD ──────────── 00001060   INPUT DATA
IEF653I SUBSTITUTION JCL - DSN=*.FT47FOO1,VOL=REF=*.FT47FOO1,DISP=(OLD,PASS),
XX   DCB=BUFNO=1                                                               00001080
XXDICTOUT DD  VOL=REF=*.FT48FOO1,DSN=*.FT48FOO1,DISP=(OLD,PASS),               00001100
XX   DCB=BUFNO=1                                                               00001120
XXCATAOUT DD  VOL=REF=*.FT47FQO1,DSN=*.FT47FOO1,DISP=(OLD,PASS),               00001140
XX   DCB=BUFNO=1                                                               00001160
XXFTO1FOO1 DD UNIT=SYSCA,SPACE=(TRK,(50,10)),DCB=(RECFM=F,BLKSIZE=80)          00001180
XXSYSIN  CD DSN=*.FTO1FQO1,DISP=(OLD,DELETE),VOL=REF=*.FTO1FOO1                00001200
//SETUP   DD  *
//

IEF283I   SYS73025.T145852.RFOCC.MO12767.RO000376   NOT DELETED 8
IEF283I   VOL SER NOS= KLOSOS 1.
IEF283I   SYS73025.T145852.RFOCC.MO12767.RO000373   NOT DELETED 8
IEF283I   VOL SER NOS= ISRB  1.
IEF283I   SYS73025.T145852.RFOCC.MO12767.RO000379   NOT DELETED 8
IEF283I   VOL SER NOS= ISRA  1.
IEF283I   SYS73025.T145852.RFOCC.MO12767.RO000380   NOT DELETED 8
IEF283I   VOL SER NOS= ISRB  1.
IEF283I   SYS73025.T145852.RFOCC.MO12767.RO0003R2   NOT DELETED 8
IEF283I   VOL SER NOS= ISRA  1.
IEF283I   SYS73025.T145852.RFOCC.MO12767.RO000383   NOT DELETED 8
IEF283I   VOL SER NOS= KLOSOS 1.
IEF283I   SYS73025.T145852.RFOCC.MO12767.RO000381   NOT DELETED 8
IEF283I   VOL SER NOS= MFT1  1.
IEF283I   SYS73025.T145852.RFOCC.MO12767.RO000370   NOT DELETED 8
IEF283I   VOL SER NOS= MFT1  1.
IEF283I   SYS73025.T145852.RFOCC.MO12767.RO0003R4   NOT DELETED 8
IEF283I   VOL SER NOS= KLOSOS 1.
ISRO11I STEP GO       EXECUTION TIME =   014.02 SEC.
ISRO13I STEP GO       CORE USAGE =   44K HSC
ISRO12I TOT. MO12767  EXECUTICN TIME =   014.02 SEC.
ISRO16I TIME OF DAY = 16.58.20, DATE = 73.025
```

```
         INSTITUTE FOR SOCIAL RESEARCH MONITOR SYSTEM    7/1/72
*****    **********************************************************************
*****    NEW DSLIST,HICLSTR,FRUILD,GSCORE,FILECCPY,REBILD,AND NTILE ON 360/40 AUGUST 21    *****
*****    SEE WHIPS 89 AND 90 FOR CHANGES.  THESE CHANGES APPLY TO OSIRIS III AND REPLACE    *****
*****    OSIRIS/40 AND OSIRIS II PROGRAMS.  PLEAS ADJUST YOUR SETUPS ACCORDINGLY.    *****
*****    **********************************************************************    *****
                                                                                  *****
                                                                                  *****
```

*****TIME IS 16:51:25

*****LISTING OF SET-UP FOLLOWS:

```
CARD          1         2         3         4         5         6         7         8
NO.  12345678901234567890123456789012345678901234567890123456789012345678901234567890
 1   $RUN MNA
 2   PRINTOUT FCR MONOGRAPH
 3       INFILE=TEST,  BADDATA=SKIP*
 4   V254,V169,V317,V403,V277*
 5   EXCLUDE V254=9 OR  V169=9 OR V317=9*
 6   REGION , RACE AND EDUCATION PREDICTING VIOLENCE TYPOLOGY  GLOBAL FILTER, PART OF OSIRIS SOFTWARE SYSTEM
 7   DEPVAR=403 WEIGHT=277 MDOPT=BOTH NAMES *
 8   PACIFISTINTERMEDWARRIORSANARCHISVIGILANT
 9   V254,V169,V317*
10   $RECODE
11       V403 = BRAC(V403,0-1=1,2=2,3-4=3,5-6=4,7-8=5,9=9)  ⎫ RECODING VARIABLES IS NOT PART OF MNA
12       V254 = BRAC(V254,0=0,1=1,2=2,3=3,7=3,4=4,6=4,8=8,9=9)  ⎬ BUT AN OPTIONAL FEATURE OF THE OSIRIS
13       V169 = BRAC(V169,1=1,2=2,3-9=9)                      ⎭ SOFTWARE SYSTEM
14       V317 = BRAC(V317,1=1,2-3=2,4=4,5-6=5,7=7,9=9)
```

RECODE   OSIRIS RECODING ROUTINE   7/1/72

'RECODE SCANNER BEGINS- 16 51 38   0

RECODE STATEMENTS:

```
 1   V403 = BRAC(V403,0-1=1,2=2,3-4=3,5-6=4,7-8=5,9=9)
 2   V254 = BRAC(V254,0=0,1=1,2=2,3=3,7=3,4=4,6=4,8=8,9=9)
 3   V169 = BRAC(V169,1=1,2=2,3-9=9)
 4   V317 = BRAC(V317,1=1,2-3=2,4=4,5-6=5,7=7,9=9)
```

RECODE SCANNER ENDS - 16 51 47   60

MNA: OSIRIS MULTIVARIATE NOMINAL ANALYSIS PROGRAM - JULY 1, 1972

PRINTOUT FOR MONOGRAPH

INFILE=TEST, BADDATA=SKIP*

THE VARIABLE LIST IS:

V254,V169,V317,V403,V277*

	VAR.	TYPE	VARIABLE NAME	TLOC	WIDTH	NODEC	RESP.	MDCODE1	MDCODE2	REFNO	ID	TSEQNO
T	169	0	RACE OF R	178	1	0	1				Z2	
T	254	0	ID REGION OF CHILDHOOD	329	1	0	1		0000009		P5	
T	277	0	SAMPLING WEIGHT	365	2	0	1					
T	317	0	R'S EDUCATION RECODED	435	1	0	1		0000009			
T	403	0	VIOLENCE TYPOLOGY	537	1	0	1		0000009			

{ DICTIONARY LISTING INFORMATION ABOUT EACH VARIABLE }

1374 DATA CASES INCLUDED AFTER GLOBAL FILTER

REGION , RACE AND EDUCATION PREDICTING VIOLENCE TYPOLOGY

DEPVAR=403 WEIGHT=277 MDOPT=BOTH NAMES *

THE VARIABLE LIST IS:

V254,V169,V317*

REGION , RACE AND EDUCATION PREDICTING VIOLENCE TYPOLOGY

PARAMETER SPECIFICATIONS

```
    DEPENDENT VARIABLE      403
    WEIGHT VARIABLE         277
    MISSING DATA OPTION     BOTH
    LOCAL FILTER OPTION     YES
    PRINT MATRIX OPTION     NO


NUMBER OF DATA CASES =     1236
SUM OF WEIGHTS =          13185.
    62 CASES FILTERED
    48 MISSING DATA CASES
    28 CASES OUTSIDE VALID CODE RANGES
```

PREDICTOR	NON-EMPTY CODES
1. V 254 ID REGION OF CHILDHOOD	0 1 2 3 4 8
2. V 169 RACE OF R	1 2
3. V 317 R'S EDUCATION RECODED	1 2 4 5 7

```
*****TIME 16 53 53 38

*****TIME 16 54 39 45
*** COLUMN 1 HAS  4. SIGNIFICANT DIGITS
*** COLUMN 2 HAS  4. SIGNIFICANT DIGITS   }  LISTINGS OF ACHIEVED PRECISION IN
*** COLUMN 3 HAS  4. SIGNIFICANT DIGITS   }  SOLUTION OF LINEAR SYSTEM
*** COLUMN 4 HAS  4. SIGNIFICANT DIGITS
*** COLUMN 5 HAS  4. SIGNIFICANT DIGITS

*****TIME 16 54 54 97
```

DEPENDENT VARIABLE V 403 VIOLENCE TYPOLOGY

	CODE	N	W	PERCENT
PACIFIST	1	287	3300.	25.03
INTERMED	2	180	2035.	15.43
WARRIORS	3	214	2035.	15.43
ANARCHIS	4	212	1900.	14.41
VIGILANT	5	343	3915.	29.69

V 254.1C REGION OF CHILDHOOD

CODE		Y	PACIFIST 1	INTERMED 2	WARRIORS 3	ANARCHIS 4	VIGILANT 5
0	N 69	PERCENT	25.61	13.41	10.98	12.20	37.80
	SUM W 820.	ADJ PCT	25.66	12.76	12.13	14.02	35.44
	PCT 6.22	COEFF	0.63	-2.68	-3.31	-0.39	5.74
1	N 220	PERCENT	30.69	16.63	13.47	11.09	29.12
	SUM W 2525.	ADJ PCT	30.23	16.51	14.30	11.85	27.11
	PCT 19.15	COEFF	5.20	1.08	-1.14	-2.56	-2.58
2	N 250	PERCENT	22.34	15.02	14.84	18.50	29.30
	SUM W 2730.	ADJ PCT	21.26	14.88	16.01	19.53	28.32
	PCT 20.71	COEFF	-3.77	-0.56	0.58	5.12	-1.38
3	N 166	PERCENT	34.24	13.40	8.68	12.90	30.77
	SUM W 2015.	ADJ PCT	32.81	12.96	10.95	14.99	28.30
	PCT 15.28	COEFF	7.78	-2.47	-4.49	0.58	-1.40
4	N 458	PERCENT	17.49	17.13	22.07	15.68	27.62
	SUM W 4145.	ADJ PCT	19.25	17.85	19.06	12.50	31.34
	PCT 31.44	COEFF	-5.78	2.42	3.62	-1.91	1.65
8	N 73	PERCENT	30.53	12.11	11.58	11.05	34.74
	SUM W 950.	ADJ PCT	30.17	11.18	13.38	13.95	31.31
	PCT 7.21	COEFF	5.15	-4.25	-2.05	-0.46	1.62

V 169.RACE OF R

CODE		Y	PACIFIST 1	INTERMED 2	WARRIORS 3	ANARCHIS 4	VIGILANT 5
1	N 949	PERCENT	26.13	16.13	13.37	11.60	32.77
	SUM W 11595.	ADJ PCT	25.31	16.47	13.88	11.34	33.00
	PCT 87.94	COEFF	0.28	1.04	-1.55	-3.07	3.31
2	N 287	PERCENT	16.98	10.38	30.50	34.91	7.23
	SUM W 1590.	ADJ PCT	23.00	7.86	26.76	36.82	5.57
	PCT 12.06	COEFF	-2.03	-7.58	11.32	22.41	-24.12

V 317.R'S EDUCATION RECODED

CODE	Y		1 PACIFIST	2 INTERMED	3 WARRIORS	4 ANARCHIS	5 VIGILANT
1	N 70	PERCENT	15.00	19.17	32.50	10.83	22.50
	SUM W 600.	ADJ PCT	19.84	19.05	27.32	6.91	26.88
	PCT 4.55	COEFF	-5.19	3.62	11.88	-7.50	-2.81
2	N 423	PERCENT	17.90	17.11	17.11	15.44	32.44
	SUM W 4470.	ADJ PCT	18.27	17.22	16.38	14.82	33.31
	PCT 33.90	COEFF	-6.76	1.79	0.94	0.41	3.62
4	N 393	PERCENT	26.96	14.73	16.75	11.64	29.93
	SUM W 4210.	ADJ PCT	26.43	14.70	17.39	12.20	29.27
	PCT 31.93	COEFF	1.40	-0.73	1.96	-2.21	-0.42
5	N 299	PERCENT	30.36	14.65	10.12	15.86	29.00
	SUM W 3310.	ADJ PCT	29.99	14.71	10.84	16.19	28.28
	PCT 25.10	COEFF	4.97	-0.73	-4.60	1.77	-1.42
7	N 51	PERCENT	45.38	8.40	5.88	21.85	18.49
	SUM W 595.	ADJ PCT	43.49	7.58	8.11	24.63	16.19
	PCT 4.51	COEFF	18.46	-7.86	-7.33	10.22	-13.50

*****TIME 16 54 59 60

REGION, RACE AND EDUCATION PREDICTING VIOLENCE TYPOLOGY

5 CODES FOR DEPENDENT VARIABLE V 403 VIOLENCE TYPOLOGY

CODE	1 PACIFIST	2 INTERMED	3 WARRIORS	4 ANARCHIS	5 VIGILANT
N	287	180	214	212	343
SUM WT	3300.	2035.	2035.	1900.	3915.
PERCENT	25.03	15.43	15.43	14.41	29.69
R-SQUARED	0.0426	0.0103	0.0439	0.0613	0.0425
R-SQUARED (ADJUSTED)	0.0347	0.0022	0.0361	0.0536	0.0347

V 254 10 REGION OF CHILDHOOD   Y

	PACIFIST	INTERMED	WARRIORS	ANARCHIS	VIGILANT
ETA-SQUARED =	0.0217	0.0022	0.0183	0.0061	0.0038
BETA-SQUARED =	0.0159	0.0037	0.0065	0.0064	0.0024

GENERALIZED ETA-SQUARE = 0.0106

BIVARIATE THETA = 0.3072

V 169 RACE OF R   Y

	PACIFIST	INTERMED	WARRIORS	ANARCHIS	VIGILANT
ETA-SQUARED =	0.0047	0.0027	0.0239	0.0467	0.0331
BETA-SQUARED =	0.0003	0.0060	0.0135	0.0558	0.0382

GENERALIZED ETA-SQUARE = 0.0218

BIVARIATE THETA = 0.3303

V 317 R'S EDUCATION RECODED   Y

	PACIFIST	INTERMED	WARRIORS	ANARCHIS	VIGILANT
ETA-SQUARED =	0.0260	0.0032	0.0199	0.0052	0.0051
BETA-SQUARED =	0.0207	0.0037	0.0120	0.0078	0.0065

GENERALIZED ETA-SQUARE = 0.0123

BIVARIATE THETA = 0.3170

****

MULTIVARIATE STATISTICS

GENERALIZED R**2   0.0402

MULTIVARIATE THETA   0.3644

CORRECTLY CLASSED WT. N	1340	0	55	530	2880
CORRECTLY CLASSED PROPORTION	0.4061	0.0	0.0270	0.2789	0.7356

CLASSIFICATION MATRIX

ACTUAL	PREDICTED					
	1 PACIFIST	2 INTERMED	3 WARRIORS	4 ANARCHIS	5 VIGILANT	
PACIFIST 1	1340	0	50	220	1690	3300
PERCENT	40.61	0.0	1.52	6.67	51.21	
INTERMED 2	460	0	35	130	1410	2035
PERCENT	22.60	0.0	1.72	6.39	69.29	
WARRIORS 3	270	0	55	430	1280	2035
PERCENT	13.27	0.0	2.70	21.13	62.90	
ANARCHIS 4	425	0	25	530	920	1900
PERCENT	22.37	0.0	1.32	27.89	48.42	
VIGILANT 5	920	0	25	90	2880	3915
PERCENT	23.50	0.0	0.64	2.30	73.56	
TOTAL	3415	0	190	1400	8180	13185

*****TIME 16 56 2 40

END OF FILE - NORMAL TERMINATION

*****TIME IS 16:56: 8

NO MORE RUN CARDS IN SETUP. STEP TERMINATED

APPENDIX D

OBTAINING THE PROGRAM AND ADAPTING IT TO OTHER SYSTEMS

Obtaining Program

Information on obtaining MNA, as part of the OSIRIS III package, may be obtained from the Institute for Social Research program librarian by writing to:

> Program Librarian
> Institute for Social Research
> Room 106
> University of Michigan
> Ann Arbor, Michigan   48106

Adaptation

MNA is written to run on an IBM 360/40 under HASP/MFT with a FORTRAN G level compiler, IBM basic ASSEMBLER and 104K available byte partition. If the adaptation is to another IBM machine with the same language software, at least 104K core and sufficient peripheral devices (tape drive and 2314 disk unit or equivalent), no problems should be encountered if the instructions given with the package are followed carefully. Listed below are common adaptation problems and solutions. If the user's adaptation does not fall into one or more of these categories, it is suggested that a systems programmer be consulted.

Less than 104K

The simplest change here is to reduce the dimensionality of the sums of squares-cross products matrix XMTRX (100,100) appropriately. For example, if one had only 78K core available (28K too small) then a 7K full word reduction to the dimensionality of XMTRX is sufficient, leading to XMTRX (54,54) and a corresponding reduction in the number of dummy predictor variables to 54. (See Chapter 5 - Restriction C).

FORTRAN E or Earlier Compiler

The basic problems here are non-recognition of the FORTRAN G LEVEL INTEGER * 2 type declaration and the "logical if" statement. Provided memory permits, all program INTEGER * 2 type statements may be converted to INTEGER * 4. In addition, in the OSIRIS subroutines the GETDIC argument LIST must be changed to INTEGER * 4 and GETDIC reprogrammed. All logical IFs can be easily converted to their "arithmetic if" counterparts.

Inadequate Temporary Disk Storage Space

DD name ISR01 and FT04F001 can be overriden in the JCL or changed in the program to specify tape units. This change will substantially increase the I/O time as data will be read from tape for each analysis packet. (See Appendix B: MNA Macro Flow Chart).

Non-IBM Machine

The major problem here is that almost without exception the Assembler coded subroutines GPIN and DIRECT cannot be assembled. This would necessitate either rewriting these subroutines in the available Assembler language or reprogramming the I/O and data manipulation portions of MNA.

Volumes on Research Methodology from the Institute for Social Research

MEASURES FOR PSYCHOLOGICAL ASSESSMENT: A Guide to 3,000 Original Sources and Their Applications by Ki-Taek Chun, Sidney Cobb, and John R.P. French, Jr. 1975. 688 p.

A TECHNIQUE FOR EVALUATING INTERVIEWER PERFORMANCE: A Manual for Coding and Analyzing Interviewer Behavior from Tape Recordings of Household Interviews by Charles F. Cannell, Sally A. Lawson, and Doris L. Hausser. 1975. 138 p.

DATA PROCESSING IN THE SOCIAL SCIENCES WITH OSIRIS by Judith Rattenbury and Paula Pelletier. 1974. 245 p.

A GUIDE FOR SELECTING STATISTICAL TECHNIQUES FOR ANALYZING SOCIAL SCIENCE DATA by Frank M. Andrews, Laura Klem, Terrence N. Davidson, Patrick O'Malley, and Willard L. Rodgers. 1974, second printing 1975. 36 p.

SEARCHING FOR STRUCTURE by John A. Sonquist, Elizabeth Lauh Baker, and James N. Morgan. 1971. Revised edition 1974. 236 p.

INTRODUCTION TO THE IBM 360 COMPUTER AND OS/JCL (JOB CONTROL LANGUAGE) by Judith Rattenbury. 1971. Revised edition 1974. 103 p.

MULTIPLE CLASSIFICATION ANALYSIS: A Report on a Computer Program for Multiple Regression Using Categorical Predictors by Frank M. Andrews, James N. Morgan, John A. Sonquist, and Laura Klem. 1967. Revised edition 1973. 105 p.

MEASURES OF SOCIAL PSYCHOLOGICAL ATTITUDES by John P. Robinson and Phillip R. Shaver. 1969. Revised edition 1973. 750 p.

MEASURES OF OCCUPATIONAL ATTITUDES AND OCCUPATIONAL CHARACTERISTICS by John P. Robinson, Robert Athanasiou, and Kendra B. Head. 1969, sixth printing 1974. 480 p.

MULTIVARIATE NOMINAL SCALE ANALYSIS: A Report on a New Analysis Technique and a Computer Program by Frank M. Andrews and Robert C. Messenger. 1973, third printing 1975. 114 p.

THAID: A Sequential Analysis Program for the Analysis of Nominal Scale Dependent Variables by James N. Morgan and Robert C. Messenger. 1973, second printing 1974. 98 p.

OSIRIS: Architecture and Design by Judith Rattenbury and Neal Van Eck. 1973, second printing 1974. 315 p.

ECONOMIC SURVEY METHODS by John B. Lansing and James N. Morgan. 1971, fourth printing 1974. 448 p.

INFERENCE FROM SURVEY SAMPLES: An Empirical Investigation by Martin R. Frankel. 1971, fourth printing 1974. 173 p.

MULTIVARIATE MODEL BUILDING: The Validation of a Search Strategy by John A. Sonquist. 1970, third printing 1975. 264 p.

The above listed volumes are published by the Institute for Social Research. For information about prices and available editions write to: Sales Fulfillment Section, Institute for Social Research, The University of Michigan, Box 1248, Ann Arbor, Michigan 48106.